THE ISSU

COLLEC

GILLDA LEITENBERG

Justice

EDITED BY MICKI CLEMENS

◆ ◆ ◆

To my two sisters, Mardi and Marlene,
who taught me that three sisters, together,
can learn what's fair and just

McGraw-Hill Ryerson Limited

Toronto • Montreal • New York • Auckland • Bogotá
Caracas • Lisbon • London • Madrid • Mexico • Milan
New Delhi • Paris • San Juan • Singapore • Sydney • Tokyo

Justice
The Issues Collection

Copyright © McGraw-Hill Ryerson Limited, 1994

ISBN 0-07-551519-9

5 6 7 8 9 10 BBM 3 2 1 0 9

Printed and bound in Canada

Canadian Cataloguing in Publication Data

Main entry under title:

Justice

(The issues collection)
For use in grades 7, 8, 9
ISBN 0-07-551519-9

1. Readers (Elementary) 2. Readers (Secondary).
3. Readers — Justice. 4. Justice — Literary
collections. I. Clemens, Micki, date. II. Series

PE1121.J87 1994   808'.0427   C94-930339-9

Editor: *Kathy Evans*
Senior Supervising Editor: *Carol Altilia*
Permissions Editor: *Jacqueline Donovan*
Copy Editor: *Judith Kalman*
Designer: *Mary Opper*
Typesetter: *Pages Design Ltd.*
Photo Researcher: *Elaine Freedman*
Cover Illustrator: *Yves LeFevre / The Image Bank Canada*

Contents

Introduction

The word "justice" is not a word that you generally introduce in your normal, everyday conversation while riding the bus to school or lining up for lunch in the cafeteria. The word possesses a certain weight and substance that is a little intimidating. Justice! Even to say the word, you have to slow down and shape the syllables carefully with your tongue and teeth. J-U-S-T-I-C-E!

When thinking about justice, you might initially have associations and images such as robed and solemn judges, lawyers, unintelligible language, episodes of courtroom dramas, last-minute confessions, abject prisoners, and fortress-like penitentiaries.

These aspects of justice might sound as if they are far removed from your life. But, the fascinating fact is that issues of justice are all around us. They are cloaked in terms like truth, fair play, suffering, balance, rewards, punishment, rules, obligations, and equality. Justice, then, is a complex concept that plays a powerful role both in formal ways through the justice system and in the guise of these more familiar issues.

On the global level, you might debate questions such as: Is it fair or just that innocent children suffer starvation or are killed in foreign conflicts? Is it fair or just that people are discriminated against because of their race, colour, or religion? Is it fair or just that women receive less pay than men, for work of equal value? Is it fair or just that one parent is refused access to his or her child in a divorce settlement?

On a more personal level, you might want to reflect on issues that touch closer to home. Have you ever felt that you were being treated unfairly by your parents? Perhaps you were grounded for a week because of a fight your sister started. How did this make you feel? Did you act on those feelings? Did you talk the situation over with anyone? Did you ever resolve the feelings you had?

And what if the shoe were on the other foot? Have you ever watched someone else being treated unfairly or unjustly? Perhaps you have seen teenagers being harassed without cause by security officers in a mall. What were your feelings in this case? Were they as strong as your own personal feelings when you were the object of the injustice? Did you act in any way to support those unfairly accused? Should you have?

In this book you will explore broadly the kinds of meaning that justice has. Reading the stories, poems, non-fiction selections, and dramatic selections will allow you to check out the way that other individuals react and respond to personal instances of wrongful action. Ideas of fairness, right conduct, guilty feelings, revenge, and, yes, forgiveness, are played out by the characters and the events in these selections.

In addition, with selections set in the more global contexts, you will explore situations relating to racial and political injustice, gender equality, laws and rules in society, and the justice system itself, as experienced by young offenders.

All of the selections are offered with the purpose of creating opportunities for you to raise questions, examine your own ideas, discuss your opinions, and find expression for your ideas to clarify and shape your own thinking on these matters.

Determining what is just and fair and whether or not you have a more active part to play in your own life and the life of your community with regard to justice for all is a worthy goal.

As a nation, we have embraced the idea that a just and fair society is one that shows concern for all people now and for future generations. By reading and reflecting on these selections, you are taking a step toward acquiring a better understanding of justice, both in its lofty forms and in its more commonplace examples.

You will be better prepared and more informed to act and react in thoughtful and insightful ways and to participate responsibly in contributing to the "just society."

I hope you are engaged, challenged, provoked, and inspired by this book.

Micki Clemens

This Is a Law

♦♦♦

BY

F.R. SCOTT

Who says Go
When the Green says Go
And who says No
When the Red says No?
Asked I.

I, said the Law,
I say Go
When the Green says Go
And don't you Go
When the Red says No,
Said the Law.

Who are you
To tell me so
To tell me Go
When the Green says Go
And tell me No
When the Red says No?
Asked I.

I am you
Said the Law.

Are you me
As I want to be?
I don't even know
Who you are.

I speak for you
Said the Law.

You speak for me?
Who told you you should?
Who told you you could?
How can this thing be
When I'm not the same as before?

I was made for you
I am made by you
I am human too
So change me if you will
Change the Green to Red
Shoot the ruling class
Stand me on my head
I will not be dead
I'll be telling you Go
I'll be telling you No
For this is a Law
Said the Law.

Forgive My Guilt

♦♦♦

BY

ROBERT P. TRISTRAM COFFIN

Not always sure what things called sins may be,
I am sure of one sin I have done.
It was years ago, and I was a boy,
I lay in the frostflowers with a gun,
The air ran blue as the flowers, I held my breath,
Two birds on golden legs slim as dream things
Ran like quicksilver on the golden sand,
My gun went off, they ran with broken wings
Into the sea, I ran to fetch them in,
But they swam with their heads high out to sea,
They cried like two sorrowful high flutes,
With jagged ivory bones where wings should be.

For days I heard them when I walked that headland
Crying out to their kind in the blue,
The other plovers were going over south
On silver wings leaving these broken two.
The cries went out one day; but I still hear them
Over all the sounds of sorrow in war or peace
I ever have heard, time cannot drown them,
Those slender flutes of sorrow never cease.
Two airy things forever denied the air!
I never knew how their lives at last were spilt,
But I have hoped for years all that is wild,
Airy, and beautiful will forgive my guilt.

My Guilt

◆◆◆

BY

AGNES

COPITHORNE

When I was a child I walked two miles to school
accompanied by a neighbour boy two years younger.
Freddie was fat and freckle faced
with wheat-straw hair and a mean stepmother.
It was late fall and one day his father
bought him a new winter cap.
It was made of heavy brown tweed with ear flaps.
It cost one dollar and twenty cents.

The next morning when he joined me
on the way to school, he showed me his new cap.
Almost bursting with pride, he took it off
so I could see the rabbit fur lining the ear flaps.
Whether out of downright meanness, or jealousy
because I didn't have a new cap,
just last year's old red knitted toque,
I snatched it out of his hand
tossing it in the air and catching it again.
This went on for about half a mile.

Screaming and pleading, he ran after me.
But his legs were shorter than mine,
he couldn't catch me.
Winded, I stopped and thrust the cap down a badger hole.
He ran up sobbing and reached down into the earth.
But the hole was deep, his arm not long enough.
He sat back on his heels and cried bitterly.
Guiltily, I stretched my arm down,
But there was no bottom, or so it appeared.
"Come on," I said, "We'll be late for school,
we'll get it on our way home tonight."

All day I felt his troubled gaze upon me
and I had trouble focusing on the printed page.
When we trudged homeward after school,
we tried again to rescue the cap, with no success.
And since Freddie was not allowed to loiter,
nor was I, we gave up.
Freddie dragged his feet, dreading to face
his stepmother and I too cowardly
to confess my guilt.

Later that evening his father took a shovel
and dug, but the hole was deep, slanting off
in different directions underground.
He gave up too and Freddie cried himself to sleep.
After that he came to school bareheaded.
My heart was like a stone in my breast
when I looked at his ears red with cold.
But I had no money to buy him another cap
even had I wanted to, which I suppose I didn't.

They moved away after that, not because of the cap,
but drought, poverty and all that goes with it
drove them to another part of the country.
Through the many years since,
Freddie's sad face haunts me accusingly
and rightly so, for the callous thing I had done,
when I was twelve and he was ten.

Groceries

♦♦♦

BY

MONTY

REID

The packages slide
past: cellophane and tin,
and the teller pokes the register
the way she'd poke a fat
man in the ribs. You're
missing one; she totals
up our bones.

I remember peaches.
I took one off the store
display and ate it as we
filled our cart.
The store detectives caught
us at the checkout.
They never said much
but I could taste that peach
for days.

Guilt

♦♦♦

BY

LEONA

GOM

your mother giving you a set of dishes
and all you said was *but I move around*
so much and you can never forget
her hurt face turning away.
the best friend you accused of
flirting with your boyfriend when
all the time you knew it was him
you just couldn't face it.
the argument with your father about
not having seen his damned magazine
then finding it in your room
and never admitting it.
telling your office mate you
agreed with her motion then
voting with the others after all.

thousands of them, little knots
you can't shake loose from your memory.
it's too late now to say you're sorry.
they contract along your nerves
to consciousness, whenever you think
you are not a bad person, there
they come, little lumps of guilt
making their daily rounds,
like doctors, keeping you sick.

Win, Lose or Learn

♦♦♦

BY

PEG

KEHRET

CAST:

· ·

Four players: *Matt, Steve, Tony, Ricky*
No special setting necessary
(Matt, Steve and Tony are On Stage as scene opens.)

Matt: We could have beat them. If Coach Girard had let the first string finish the game, we would have won.

Steve: No question. We had them. But, oh no, Coach had to let everybody have their turn to play.

Matt: It's the most stupid coaching policy I've ever heard of. You don't see professional coaches making sure every player has a chance to play in every game.

Steve: I couldn't believe it when he took me out and put Ricky Schuster in. Ricky Schuster hasn't made a free throw the entire season. He just barely made the team.

Matt: Maybe Coach had a big bet on this game. Maybe he has a secret gambling problem, and he had to make sure we lost so he could collect the money and pay his debts.

Tony: Oh, come on, Matt. You know that isn't true. He just feels strongly that every person on the team should get some playing time. If you were a sub instead of being on the starting team, you'd probably like his policy.

Matt: Not if it meant we had to lose the biggest game of the season by two points. Two lousy points! And all because the worst players on the entire squad got to play the last five minutes.

Tony: Those guys come to practice every day. They work just as hard as we do.

Steve: Sure, they come to practice. But that doesn't make them good players. At least, not good enough to go in for the final five minutes of the most important game of the season.

Matt: My dad is going to have a fit. The last time we lost a game because of this dumb rule, Dad threatened to complain to the school board. This time, he'll do it. I know he will. I could hear him booing.

Steve: Maybe the school board will overrule Coach.

Matt: Or fire him.

Tony: Is that what you want? You told me Coach Girard has done more to help you improve your game than any other coach you've ever had.

Matt: I'm not saying he's a bad coach. I'm just saying he has a rotten philosophy about who gets to play. What's so important about sending Ricky Schuster in for five minutes so he can miss two free throws, double dribble, and throw the ball straight into the hands of the other team's center?

Tony: Shhh. Here comes Ricky. He'll hear you. (*He points. Matt and Steve look. Ricky enters.*)

Ricky: Hi, guys.

Others: Hi, Ricky.

Ricky: Sorry about the game.

Tony: Yeah.

Ricky: I wanted to play my best today. My grandparents came over a hundred miles just to see the game. I was really hoping I'd make those free throws.

Matt: So were we.

Tony: Maybe next time.

Matt: I doubt it.

Ricky: It's so frustrating. At home, I practice free throws, and I sink eighty percent of them. Swish! Straight through the net. Then I get in a tight game, and I don't know what happens.

Matt: I do. You miss.

Tony: Lay off, Matt. He feels bad enough.

Ricky: At least I got in for a few minutes. I would have really felt terrible if Grandpa and Grandma came so far to watch me play and then I never even got in the game.

Matt: Don't they care that your team lost?

Ricky: Well, naturally, they'd like to see us win. But they agree with Coach Girard's philosophy that the important thing is for each player to try to play up to his best potential. If you win, that's icing on the cake.

Steve: Was today your best potential?

Ricky: No way. Coach says I need more experience. He says my coordination was slow to develop, but if I keep practicing, I'll be a good player someday.

Matt: Someday. What happens in the meantime?

Steve: It's great to get practice, but does it have to come when we're tied with five minutes to go?

Tony: Hey! He said he was sorry about the game.

Ricky: I need practice under pressure, too. So do the other subs. We have to know how it feels.

Matt: Well, I'd like to know how it feels to win a game.

Tony: We've won some games. Six wins, four losses.

Steve: It could have been eight wins and two losses.

Matt: My dad is really going to chew me out tonight.

Ricky: For what? You played a great game.

Matt: Most of the time. But I missed both of those long shots, the three-pointers. If I'd made either one of them, we would have won.

Tony: You aren't the only person who missed a shot.

Ricky: That's for sure.

Matt: I guess you'll really get yelled at when you get home.

Ricky: No. If I do something good, my folks tell me it was great. If I play lousy, like today, they just say it was fun to come and watch the game.

Tony: That's how my mom is, too. She says as long as I play clean, she's proud of me.

Ricky: Well, I have to get going. My grandparents are taking the whole family out for pizza. See you at practice tomorrow. (*Ricky exits.*)

Matt: Unfortunately.

The Mouth-organ Boys

◆ ◆ ◆

BY

JAMES

BERRY

I wanted a mouth-organ, I wanted it more than anything else in the whole world. I told my mother. She kept ignoring me but I still wanted a mouth-organ badly.

I was only a boy. I didn't have a proper job. Going to school was like a job, but nobody paid me to go to school. Again I had to say to my mother, "Mum, will you please buy a mouth-organ for me?"

It was the first time now, that my mother stood and answered me properly. Yet listen to what my mother said. "What d'you want a mouth-organ for?"

"All the other boys have a mouth-organ, mam," I told her.

"Why is that so important? You don't have to have something just because others have it."

"They won't have me with them without a mouth-organ, mam," I said.

"They'll soon change their minds, Delroy."

"They won't, mam. They really won't. You don't know Wildo Harris. He never changes his mind. And he never lets any other boy change his mind either."

"Delroy, I haven't got the time to argue with you. There's no money to buy a mouth-organ. I bought you new shoes and clothes for Independence Celebrations. Remember?"

"Yes, mam."

"Well, money doesn't come on trees."

"No, mam." I had to agree.

"It's school-day. The sun won't stand still for you. Go and feed the fowls. Afterwards milk the goat. Then get yourself ready for school."

She sent me off. I had to go and do my morning jobs.

Oh my mother never listened! She never understood anything. She always had reasons why she couldn't buy me something and it was no good wanting to talk to my dad. He always cleared off to work early.

All my friends had a mouth-organ, Wildo, Jim, Desmond, Len — everybody had one, except me. I couldn't go round with them now. They wouldn't let anybody go round with them without a mouth-organ. They were now "The Mouth-organ Boys." And we used to be all friends. I used to be their friend. We all used to play games together, and have fun together. Now they pushed me way.

"Delroy! Delroy!" my mother called.

I answered loudly. "Yes, mam!"

"Why are you taking so long feeding the fowls?"

"Coming, mam."

"Hurry up, Delroy."

Delroy. Delroy. Always calling Delroy!

I milked the goat. I had breakfast. I quickly brushed my teeth. I washed my face and hands and legs. No time left and my mother said nothing about getting my mouth-organ. But my mother had time to grab my head and comb and brush my hair. She had time to wipe away toothpaste from my lip with her hand. I had to pull myself away and say, "Good day, Mum."

"Have a good day, Delroy," she said, staring at me.

I ran all the way to school. I ran wondering if the Mouth-organ Boys would let me sit with them today. Yesterday they didn't sit next to me in class.

I was glad the boys came back. We all sat together as usual. But they teased me about not having a mouth-organ.

Our teacher, Mr. Goodall, started writing on the blackboard. Everybody was whispering. And it got to everybody talking quite loudly. Mr. Goodall could be really cross. Mr. Goodall had big muscles. He had a moustache too. I would like to be like Mr. Goodall when I grew up. But he could be really cross. Suddenly Mr. Goodall turned round and all the talking stopped, except for the voice of Wildo Harris. Mr. Goodall held the chalk in his hand and stared at Wildo Harris. He looked at Teacher and dried up. The whole class giggled.

Mr. Goodall picked out Wildo Harris for a question. He stayed sitting and answered.

"Will you please stand up when you answer a question?" Mr. Goodall said.

Wildo stood up and answered again. Mr. Goodall ignored him and asked another question. Nobody answered. Mr. Goodall pointed at me and called my name. I didn't know why he picked on me. I didn't know I knew the answer. I wanted to stand up slowly, to kill time. But I was there, standing. I gave an answer.

"That is correct," Mr. Goodall said.

I sat down. My forehead felt hot and sweaty, but I felt good. Then in schoolyard at recess time, Wildo joked about it. Listen to what he had to say: "Delroy Brown isn't only a big head. Delroy Brown can answer questions with big mouth."

"Yeh!" the gang roared, to tease me.

Then Wildo had to say, "If only he could get a *mouth*-organ." All the boys laughed and walked away.

I went home to lunch and as usual I came back quickly. Wildo and Jim and Desmond and Len were together, at the bench, under the palm tree. I went up to them. They were swapping mouth-organs, trying out each one. Everybody made sounds on each mouth-organ, and said something. I begged Len, I begged Desmond, I begged Jim, to let me try out their mouth-organs. I only wanted a blow. They just carried on making silly sounds on each other's mouth-organs. I begged Wildo to lend me his. He didn't even look at me.

I faced Wildo. I said, "Look. I can do something different as a Mouth-organ Boy. Will you let me do something different?"

Boy, everybody was interested. Everybody looked at me.

"What different?" Wildo asked.

"I can play the comb," I said.

"Oh, yeh," Wildo said slowly.

"Want to hear it?" I asked. "My dad taught me how to play it."

"Yeh," Wildo said. "Let's hear it." And not one boy smiled or anything. They just waited.

I took out my comb. I put my piece of tissue paper over it. I began to blow a tune on my comb and had to stop. The boys were laughing too much. They laughed so much they staggered about. Other children came up and laughed too. It was all silly, laughing at me.

I became angry. Anybody would get mad. I told them they could keep their silly Mouth-organ Boys business. I told them it only happened because Desmond's granny gave him a mouth-organ for his birthday. And it only caught on because Wildo went and got a mouth-organ too. I didn't sit with the boys in class that afternoon. I didn't care what the boys did.

I went home. I looked after my goats. Then I ate. I told my mum I was going for a walk. I went into the centre of town where I had a great surprise.

The boys were playing mouth-organs and dancing. They played and danced in the town square. Lots of kids followed the boys and danced around them.

It was great. All four boys had the name "The Mouth-organ Boys" across their chests. It seemed they did the name themselves. They cut out big coloured letters for the words from newspapers and magazines. They gummed the letters down on a strip of brown paper, then they made a hole at each end of the paper. Next a string was pushed through the holes, so they could tie the names round them. The boys looked great. What a super name: "The Mouth-organ Boys"! How could they do it without me!

"Hey, boys!" I shouted, and waved. "Hey, boys!" They saw me. They jumped up more with a bigger act, but ignored me. I couldn't believe Wildo, Jim, Desmond and Len enjoyed themselves so much and didn't care about me.

I was sad, but I didn't follow them. I hung about the garden railings, watching. Suddenly I didn't want to watch any more. I went home slowly. It made me sick how I didn't have a mouth-organ. I didn't want to eat. I didn't want the lemonade and bun my mum gave me. I went to bed.

Mum thought I wasn't well. She came to see me. I didn't want any fussing about. I shut my eyes quickly. She didn't want to disturb me. She left me alone. I opened my eyes again.

If I could drive a truck I could buy loads of mouth-organs. If I was a fisherman I could buy a hundred mouth-organs. If I was an aeroplane pilot I could buy truck-loads of mouth-organs. I was thinking all those things and didn't know when I fell asleep.

Next day at school The Mouth-organ Boys sat with me. I didn't know why but we just sat together and joked a little bit. I felt good running home to lunch in the usual bright sunlight.

I ran back to school. The Mouth-organ Boys were under the palm tree, on the bench. I was really happy. They were really unhappy and cross and this was very strange.

Wildo grabbed me and held me tight. "You thief!" he said.

The other boys came around me. "Let's search him," they said.

"No, no!" I said. "No."

"I've lost my mouth-organ and you have stolen it," Wildo said.

"No," I said. "No."

"What's bulging in your pocket, then?"

"It's mine," I told them. "It's mine."

The boys held me. They took the mouth-organ from my pocket.

"It's mine," I said. But I saw myself up to Headmaster. I saw myself getting caned. I saw myself disgraced.

Wildo held up the mouth-organ. "Isn't this red mouth-organ mine?"

"Of course it is," the boys said.

"It's mine," I said. "I got it at lunchtime."

"Just at the right time, eh?," Desmond said.

"Say you borrowed it," Jim said.

"Say you were going to give it back," Len said.

Oh, I had to get a mouth-organ just when Wildo lost his! "My mother gave it to me at lunchtime," I said.

"Well, come and tell Teacher," Wildo said.

Bell rang. We hurried to our class. My head was aching. My hands were sweating. My mother would have to come to school, and I hated that.

Wildo told our teacher I stole his mouth-organ. It was no good telling Teacher it was mine, but I did. Wildo said his mouth-organ was exactly like that. And I didn't have a mouth-organ.

Mr. Goodall went to his desk. And Mr. Goodall brought back Wildo's grubby red mouth-organ. He said it was found on the floor.

How could Wildo compare his dirty red mouth-organ with my new, my beautiful, my shining clean mouth-organ? Mr. Goodall made Wildo Harris say he was sorry.

Oh it was good. It was good to become one of "The Mouth-organ Boys."

Maria Preosi

♦♦♦

BY

MEL

GLENN

I was always known as "Tracy's sister,"
Even by my friends.
In the unspoken competition between us
I was always judged by her blond hair and giggly
laugh.
I lost the race at the starting line.
For years I sat at her desk, two years behind,
Plotting.
Anger churned within me, an oil-black rage.
I wanted to kill for every injustice ever done to me.
Yet I accepted calmly, meekly,
My position in her shadow
And did not even whisper a syllable of revenge.

The School Yard Bully

◆◆◆

BY

PEG

KEHRET

Andrew Buckingham is a bully. He's mean to younger kids, and once, when Andrew thought nobody was around, Clancy Schuman saw Andrew kick a little dog. He said when the dog yelped, Andrew laughed, and kicked it again.

If there's anything in this world I can't stand, it's a bully. I never did like Andrew Buckingham and after I heard about the dog, I just plain detested him.

The problem with bullies is that it's hard to know what to do about them. Most bullies pick out one kid at a time for their victim. Usually, it's a kid who is small for his age or who is somehow different from the other kids. Then the bully hassles that one kid mercilessly.

That's what Andrew did to Clancy. I don't know if it was because Andrew found out that Clancy saw him kick the dog and blabbed it all over school, or whether he was already picking on Clancy when the dog incident happened.

Day after day, Andrew would go up to Clancy on the school yard and insult him. If Clancy talked back, Andrew punched him. If Clancy ignored the insults, Andrew called him a chicken. Either way, Clancy

couldn't win. It got so he made up excuses not to go out during recess because he knew Andrew would be waiting for him the minute he left the building.

The rest of us watched these proceedings nervously. We felt sorry for Clancy, but we also knew that anyone who intervened would be the next victim. Nobody was eager to claim that honor.

Why didn't Clancy tell the teacher? I guess Clancy thought it was better to get punched around by Andrew than to be known as a tattle-tale. In our school, tattling was a sin even worse than bullying.

And then one day, Clancy fought back. I don't know why that particular day was any different from all the days before, but during afternoon recess, Clancy was shooting baskets when Andrew went over to him and announced that he wanted to shoot baskets. Clancy went on dribbling the ball.

"Didn't you hear me, Turkey?" Andrew said. "I said, it's my turn, so hand over the basketball."

The minute Andrew raised his voice, a crowd gathered. We all stood in a semicircle around the side of the basketball court, waiting to see what would happen.

Clancy dribbled again, lifted his arms and aimed the ball for the basket. It hit the rim and bounced away. Andrew lunged for it, but Clancy was too quick for him. He darted forward, tipped the ball away from Andrew, caught it and dribbled back to the free-throw line.

Andrew got red in the face. He swore at Clancy and said he wanted the basketball NOW.

Clancy shook his head. Andrew started toward him, his fists clenched, but Clancy stood his ground, clutching that basketball tight to his chest.

As soon as Andrew was close enough, he socked Clancy's shoulder. Clancy winced but he didn't let go of the ball. Andrew whacked him again.

My heart was pounding. I didn't want Clancy to give in — but I didn't want to watch him get beat up, either.

After Andrew punched him the second time, Clancy set the basketball on the ground next to his feet. Then, without saying a word, he swung his fist at Andrew. Andrew ducked — and Clancy missed him.

Andrew's next blow caught Clancy on the cheek — and sent him staggering backwards, away from the basketball. Andrew leaned forward to pick it up.

That's when I looked around. There were two dozen of us watching and I felt ashamed that none of us would help Clancy. I wasn't eager to have Andrew punch me out, but I knew I couldn't stand there and watch any longer.

As Andrew bent to pick up the basketball, I stepped forward and kicked it out of his reach. I kicked it toward Clancy and if I live to be a thousand, I'll never forget the gratitude in Clancy's eyes as he looked to see who had kicked that ball to him.

I didn't have long to enjoy the look, because Andrew started toward me, and I could practically see the smoke pouring out of his ears.

"So, you want in the game, do you, Twerp?" he growled.

I stood next to Clancy, my knees shaking. "Clancy had the ball," I said. "You have no right to take it away from him."

I braced myself for the blows that I was sure were coming. And then something extraordinary happened. Two other kids stepped forward, one on either side of me, and they told Andrew I was right: that it was Clancy's ball.

As soon as they did that, the rest of the kids surged forward. They gathered around Clancy and me and they all told Andrew to bug off and leave Clancy alone. It was no longer Andrew against Clancy; it was Andrew against the whole fifth grade class.

Just then, the bell rang. Recess was over and we had to go back inside. It was a relief to sit at my desk. I had been certain I was going to be punched to a pulp and left to die on the basketball court.

Our collective triumph over Andrew exhilarated me but I kept wondering why we didn't stand up to him sooner. If all of us had stepped in to defend Clancy the first time Andrew picked on him, the whole sorry situation would never have happened. All it took was one show of unity to stop a bully; so why didn't one of us ever suggest to the others that we do it? Why did we let Andrew pick on Clancy all those months?

I have no answers. I can only say that I'm glad we finally put a stop to it. Recess is a lot more fun now.

Blessed
Are The
Meek

◆◆◆

BY

A.P. CAMPBELL

"**T**hat's Fred MacLean comin' up over the hill there now. I can tell the car from here — and the dust he raises." Standing in the doorway of his general store, Jim O'Brien pointed toward the distant green hillside, up past the church, where the red clay road escaped from the confines of the village, sauntered off lazily toward the horizon, like a farm boy coming home from school. There were not very many cars around that section of the coast, and Jim knew them all by sight; he made it a point of honour to recognize them before they arrived at the church. People told Jim he just kept his store there for the sake of seeing the view, and knowing what went on...."Guess they're right at that," he said chuckling. Turning his eyes to the right, he could see the fishing shacks at the harbour, and the boats putting out the trawls. Old Ben there in his houseboat, using the old handline, still fishing at eighty-five.... Jerome Gallant, with his new boat, bringing in a catch of cod to the cooperative wharf.... Jim loved the sound of the motor boats; he loved to look at the sea.

He followed his customer, Tom Ready, into the store.

"I suppose you know Fred MacLean," he continued, just to keep up a conversation.

"No — can't say I do. I've been away all winter."

"He's got the old MacLean place out by the old mill. Fred is a brother to Frank; he died and left it to Fred. He's been here about a month now. Come down from the States to live here. Spent the last forty years in Deetroyit, workin' in a car factory. Goin' to retire — do a bit of farmin', and live on his money. They say he have plenty."

"I knew Frank well, but never knew he had a brother. Never saw him here at all. So Frank left him the place? People all thought it would be goin' to that young fellow he had there with him...what was his name?"

"Gene Allan. Yes, sort of a surprise. MacLeans were all funny like that. That family, I mean. Hated to let the land pass out o' the family."

"Was Allan related to him?"

"No. Got him from the orphanage. Never really adopted him. Nice young fellow. Quiet, worked hard. Never mixed much, though."

"How did Allan take it about not gettin' the place?"

"All right, I guess. I imagine he was glad to get the chance to get away some place."

"Is he gone away?"

"Yes — went yesterday. Dunno where. Maybe I can find out from Fred when he's in here."

"How does Fred like it here? He ever do any farmin' before?"

"He was born here; went away when he was a boy. Guess he likes it all right. His wife keeps complain' about the poor conveniences, they say."

"Huh! Wait till the winter comes."

"Wonder did they give Allan anything for the years he put in on the farm?"

"Dunno. Gene wouldn't say anything, of course. He was in here lots of times this spring gettin' seed and all sorts of stuff. Kind of fixin' the place up. Never even hinted he was leavin'. He was like that. Good farmer.... Too bad.... Hate to see him go. They say Fred coaxed him to stay on, but he said he was kind of tired of farmin'."

Tom Ready picked up his fig of twist from the counter and sauntered out into the early June sunlight, leaving Jim squinting out the window. He looked with curiosity at the dust-covered car, nodded to Fred MacLean and made some excuse to follow him into the store, where he examined some axe handles, one ear cocked to the conversation at the counter.

Jim's curiosity was that of a collector — he collected people, their mannerisms, their speech, their deeds and their opinions. But he never expended the energy to classify, label or judge people...he left that to others.

"Good day, Mr. MacLean. Pretty hot and dry, eh?" he was all efficient courtesy. "The missus well?"

"Yes, it is hot. All that dust! Yes, my wife is well, thank you. Do you have any canned spaghetti? No, don't bother. Give me five pounds of sugar. Pack of cigarettes."

"'Fraid I got no tailor mades, but have some Ogins roll-your-own and some papers. Lots of pipe tobacco, some cigars...."

"Give me half a dozen cigars."

As he fished the cigars out of the box, Jim kept fishing for odds and ends of Fred MacLean's affairs.

"Nice to have you folks here.... Hate to see a good farm vacant.... Knew your brother right well. Good neighbour and a good farmer.... Guess you were gone off to the States before I come over here from Mount Stuart way.... Good thing you have that young Allan lad to help out.... A natural farmer, they say." He cast his bait carefully.

"Gene Allan is not with me now," Fred told him in a matter-of-fact way, knowing that Jim already had wind of that.

"You don't say! Gone away, huh?"

"Yes...wanted to get away. Gone West. Has folks out in Alberta. I'm going to miss him." He laughed dryly, "I'm afraid whatever I knew about farming I forgot years ago."

"Ah, sure, you'll pick it up again. You never really forget things like that," said Jim kindly. "There's lots of people around ready to give you a hand.... Well, well, Gene gone just like that.... Fine young fellow."

"Yes, indeed. Well liked around here, I can see."

Tom had found an axe handle that suited him and come over to the counter to have it put on his account.

"I guess you know Tom Ready, don't you? Tom this is Fred MacLean."

"Hello, Fred. Was just lookin' at your car. Sure picks up that red dust.... How's the farmin' goin'?"

"I guess the farming is coming along. Maybe a bit slow. I gotta roll up my sleeves and get down to business. I'm alone, now."

"Yes, I heard you tell Jim that Gene Allan was gone away." Tom lacked Jim's finesse in questioning. "Too bad. Guess a lot of folks were surprised he went," he said bluntly.

Fred MacLean reddened with anger. Although he had been away for years, he still knew the ways of the people. "Guess they do. They think he should have had the farm. That's it, isn't it?"

Tom was too much surprised by this direct language to say anything for a moment; but Jim moved in quickly oiling the waters. "Heavens no! Nobody ever thought the like of that." There were ways of getting around things, thought Jim, without being so blunt. Comes of being away.

"There was no reason he should have had it," continued Fred, anxious to vindicate his position. "He was no relation to us. He was never adopted legally. He never expected to get the place. After all, he had his living all those years, and got his schooling too."

"Of course, of course," said Jim and Tom nodded vigorously.

"As a matter of fact," pursued Fred, "he *did* appreciate it all. I must say he is a wonderful young man...wish he had stayed. He wouldn't leave till he had the crops in. Did a lot of tidying up around the place, too."

"I'm sure you did," said Jim, who liked everything to be pleasant and easy. He was finding out more than he had expected. "He was in here a lot durin' the spring, buying stuff for you — seeds and all that. Never said he was leavin'...guess young fellows are like that today." He decided to cast in another direction. "Hope the missus doesn't find it too quiet around here, seein' she's from away."

Fred did not notice that bait, he was too bent on removing from the public mind any notion that Gene Allan had left him in an ill mood. "He put in that large field of potatoes all by himself, I never even had to go out to the field, I guess I wouldn't have been much help."

"How are your spuds comin' along, with this dry spell?" cut in Tom.

"Well, they don't seem to be up yet, the field looks red still. But he said they were a special kind and he gave them extra treatment. He put all that fertilizer, that he got in here, on the land, then because it was dry he used the sprayer to wet the drills and then rolled them. He said that they might be slow coming up but they would surprise me."

"Well, that's the kind of boy he is," said Jim, enthusiastically, stopping to wait on another customer who had come in. "Angie, you remember Fred MacLean, just took over his brother Frank's place. Fred, this is Angie Joe MacLellan, you should remember him."

"Yes, I guess so. We probably went to school together."

"Glad to see you, Fred. Hope you like the farmin'. Not much money in it."

Jim's mind, fingering among Fred's statements, suddenly found a discrepancy. "Fertilizer, Fred? Did you say he got fertilizer here?"

"Yes, those bags of fertilizer. He said he got it here."

"He got it here all right." Jim closed his mouth suddenly, then gave a kind of sick laugh.

"What's wrong?" Fred was puzzled by the look on Jim's face.

"I'm afraid there's been a terrible mistake. A terrible mistake, Mr. MacLean. That wasn't fertilizer — that was cement."

Oranges

♦♦♦

BY

GARY

SOTO

The first time I walked
With a girl, I was twelve,
Cold, and weighted down
With two oranges in my jacket.
December. Frost cracking
Beneath my steps, my breath
Before me, then gone,
As I walked toward
Her house, the one whose
Porch light burned yellow
Night and day, in any weather.
A dog barked at me, until
She came out pulling
At her gloves, face bright
With rouge. I smiled,
Touched her shoulder, and led
Her down the street, across
A used car lot and a line
Of newly planted trees,
Until we were breathing
Before a drugstore. We
Entered, the tiny bell
Bringing a saleslady

Down a narrow aisle of goods.
I turned to the candies
Tiered like bleachers,
And asked what she wanted —
Light in her eyes, a smile
Starting at the corners
Of her mouth. I fingered
A nickel in my pocket.
And when she lifted a chocolate
That cost a dime,
I didn't say anything.
I took the nickel from
My pocket, then an orange,
And set them quietly on
The counter. When I looked up,
The lady's eyes met mine,
And held them, knowing
Very well what it was all
About.

   Outside,
A few cars hissing past,
Fog hanging like old
Coats between the trees.
I took my girl's hand
In mine for two blocks,
Then released it to let
Her unwrap the chocolate.
I peeled my orange
That was so bright against
The gray of December
That, from some distance,
Someone might have thought
I was making a fire in my hands.

Miss Calista's Peppermint Bottle

♦♦♦

BY

L.M. MONTGOMERY

Miss Calista was perplexed. Her nephew, Caleb Cramp, who had been her right-hand man for years and whom she had got well broken into her ways, had gone to the Klondike, leaving her to fill his place with the next best man; but the next best man was slow to appear, and meanwhile Miss Calista was looking about her warily. She could afford to wait a while, for the crop was all in and the fall ploughing done, so that the need of a successor to Caleb was not as pressing as it might otherwise have been. There was no lack of applicants, such as they were. Miss Calista was known to be a kind and generous mistress, although she had her "ways," and insisted calmly and immovably upon wholehearted compliance with them. She had a small, well-cultivated farm and a comfortable house, and her hired men lived in clover. Caleb Cramp had been perfection after his kind, and Miss Calista did not expect to find his equal. Nevertheless, she set up a certain standard of requirements; and although three weeks, during which Miss Calista had been obliged to put up with the immature services of a neighbour's boy, had elapsed since Caleb's departure, no one had as yet stepped into his vacant and coveted shoes.

Certainly Miss Calista was somewhat hard to please, but she was not thinking of herself as she sat by her front window in the chilly November twilight. Instead, she was musing on the degeneration of hired men, and reflecting that it was high time the wheat was thrashed, the house banked, and sundry other duties attended to.

Ches Maybin had been up that afternoon to negotiate for the vacant place, and had offered to give satisfaction for smaller wages than Miss Calista had ever paid. But he had met with a brusque refusal, scarcely as civil as Miss Calista had bestowed on drunken Jake Stinson from the Morrisvale Road.

Not that Miss Calista had any particular prejudice against Ches Maybin, or knew anything positively to his discredit. She was simply unconsciously following the example of a world that exerts itself to keep a man down when he is down and prevent all chance of his rising. Nothing succeeds like success, and the converse of this is likewise true — that nothing fails like failure. There was not a person in Cooperstown who would not have heartily endorsed Miss Calista's refusal.

Ches Maybin was only eighteen, although he looked several years older, and although no flagrant misdoing had ever been proved against him, suspicion of such was not wanting. He came of a bad stock, people said sagely, adding that what was bred in the bone was bound to come out in the flesh. His father, old Sam Maybin, had been a shiftless and tricky rascal, as everybody knew, and had ended his days in the poorhouse. Ches's mother had died when he was a baby, and he had come up somehow, in a hand-to-mouth fashion, with all the cloud of heredity hanging over him. He was always looked at askance, and when any mischief came to light in the village, it was generally fastened on him as a convenient and handy scapegoat. He was considered sulky and lazy, and the local prophets united in predicting a bad end for him sooner or later; and, moreover, diligently endeavoured by their general treatment of him to put him in a fair way to fulfil their predictions. Miss Calista, when she had shut Chester Maybin out into the chill gloom of the November dusk, dismissed him from her thoughts. There were other things of more moment to her just then than old Sam Maybin's hopeful son.

There was nobody in the house but herself, and although this was neither alarming nor unusual, it was unusual — and Miss Calista considered it alarming — that the sum of five hundred dollars should at that very moment be in the upper right-hand drawer of the sideboard, which sum had been up to the previous day safe in the coffers of the Millageville bank. But certain unfavourable rumours were in course of circulation about that same institution, and Miss Calista, who was nothing if not prudent, had gone to the bank that very morning and withdrawn her deposit. She intended to go over to Kerrytown the very next day and deposit it in the Savings Bank there. Not another day would she keep it in the house, and, indeed, it worried her to think she must keep it even for the night, as she had told Mrs. Galloway that afternoon during a neighbourly back-yard chat.

"Not but what it's safe enough," she said, "for not a soul but you knows I've got it. But I'm not used to have so much by me, and there are always tramps going round. It worries me somehow. I wouldn't give it a thought if Caleb was here. I s'pose being all alone makes me nervous."

Miss Calista was still rather nervous when she went to bed that night, but she was a woman of sound sense and was determined not to give way to foolish fears. She locked doors and windows carefully, as was her habit, and saw that the fastenings were good and secure. The one on the dining-room window, looking out on the back yard, wasn't; in fact, it was broken altogether; but, as Miss Calista told herself, it had been broken just so for the last six years, and nobody had ever tried to get in at it yet, and it wasn't likely anyone would begin tonight.

Miss Calista went to bed and, despite her worry, slept soon and soundly. It was well on past midnight when she suddenly wakened and sat bolt upright in bed. She was not accustomed to waken in the night, and she had the impression of having been awakened by some noise. She listened breathlessly. Her room was directly over the dining-room, and an empty stovepipe hole opened up through the ceiling of the latter at the head of her bed.

There was no mistake about it. Something or some person was moving about stealthily in the room below. It wasn't the cat — Miss Calista had shut him in the woodshed before she went to bed, and he couldn't possibly get out. It must certainly be a beggar or tramp of some description.

Miss Calista might be given over to nervousness in regard to imaginary thieves, but in the presence of real danger she was cool and self-reliant. As noiselessly and swiftly as any burglar himself, Miss Calista slipped out of bed and into her clothes. Then she tip-toed out into the hall. The late moonlight, streaming in through the hall windows, was quite enough illumination for her purpose, and she got downstairs and was fairly in the open doorway of the dining-room before a sound betrayed her presence.

Standing at the sideboard, hastily ransacking the neat contents of an open drawer, stood a man's figure, dimly visible in the moonlight gloom. As Miss Calista's grim form appeared in the doorway, the midnight marauder turned with a start and then, with an inarticulate cry, sprang, not at the courageous lady, but at the open window behind him.

Miss Calista, realizing with a flash of comprehension that he was escaping her, had a woman-like impulse to get a blow in anyhow; she grasped and hurled at her unceremonious caller the first thing that came to hand — a bottle of peppermint essence that was standing on the side-board.

The missile hit the escaping thief squarely on the shoulder as he sprang out of the window, and the fragments of glass came clattering down on the sill. The next moment Miss Calista found herself alone, standing by the sideboard in a half-dazed fashion, for the whole thing had passed with such lightning-like rapidity that it almost seemed as if it were the dissolving end of a bad dream. But the open drawer and the window, where the bits of glass were glistening in the moonlight, were no dream. Miss Calista recovered herself speedily, closed the window, lit the lamp, gathered up the broken glass, and set up the chairs which the would-be thief had upset in his exit. An examination of the sideboard showed the precious five hundred safe and sound in an undisturbed drawer.

Miss Calista kept grim watch and ward there until morning, and thought the matter over exhaustively. In the end she resolved to keep her own counsel. She had no clue whatever to the thief's whereabouts or identity, and no good would come of making a fuss, which might only end in throwing suspicion on someone who might be quite innocent.

When the morning came Miss Calista lost no time in setting out for Kerrytown, where the money was soon safely deposited in the bank. She heaved a sigh of relief when she left the building.

I feel as if I could enjoy life once more, she said to herself. Goodness me, if I'd had to keep that money by me for a week itself, I'd have been a raving lunatic by the end of it.

Miss Calista had shopping to do and friends to visit in town, so that the dull autumn day was well nigh spent when she finally got back to Cooperstown and paused at the corner store to get a bundle of matches.

The store was full of men, smoking and chatting around the fire, and Miss Calista, whose pet abomination was tobacco smoke, was not at all minded to wait any longer than she could help. But Abiram Fell was attending to a previous customer, and Miss Calista sat grimly down by the counter to wait her turn.

The door opened, letting in a swirl of raw November evening wind and Ches Maybin. He nodded sullenly to Mr. Fell and passed down the store to mutter a message to a man at the further end.

Miss Calista lifted her head as he passed and sniffed the air as a charger who scents battle. The smell of tobacco was strong, and so was that of the open boxes of dried herring on the counter, but plainly, above all the common mingled odours of a country grocery, Miss Calista caught a whiff of peppermint, so strong as to leave no doubt of its origin. There had been no hint of it before Ches Maybin's entrance.

The latter did not wait long. He was out and striding along the shadowy road when Miss Calista left the store and drove smartly after him. It never took Miss Calista long to make up her mind about anything, and she had weighed and passed judgement on Ches Maybin's case while Mr. Fell was doing up her matches.

The lad glanced up furtively as she checked her fat grey pony beside him.

"Good evening, Chester," she said with brisk kindness. "I can give you a lift, if you are going my way. Jump in, quick — Dapple is a little restless."

A wave of crimson, duskily perceptible under his sunburned skin, surged over Ches Maybin's face. It almost seemed as if he were going to blurt out a blunt refusal. But Miss Calista's face was so guileless and her tone so friendly, that he thought better of it and sprang in beside her, and Dapple broke into an impatient trot down the long hill lined with its bare, wind-writhen maples.

After a few minutes' silence Miss Calista turned to her moody companion.

"Chester," she said, as tranquilly as if about to ask him the most ordinary question in the world, "why did you climb into my house last night and try to steal my money?"

Ches Maybin started convulsively, as if he meant to spring from the buggy at once, but Miss Calista's hand was on his arm in a grasp none the less firm because of its gentleness, and there was a warning gleam in her grey eyes.

"It won't mend matters trying to get clear of me, Chester. I know it was you and I want an answer — a truthful one, mind you — to my question. I am your friend, and I am not going to harm you if you tell me the truth."

Her clear and incisive gaze met and held irresistibly the boy's wavering one. The sullen obstinacy of his face relaxed.

"Well," he muttered finally, "I was just desperate, that's why. I've never done anything real bad in my life before, but people have always been down on me. I'm blamed for everything, and nobody wants anything to do with me. I'm willing to work, but I can't get a thing to do. I'm in rags and I haven't a cent, and winter's coming on. I heard you telling Mrs. Galloway yesterday about that money. I was behind the fir hedge and you didn't see me. I went away and planned it all out. I'd get in some way — and I meant to use the money to get away out west as far from here as I could, and begin life there, where nobody knew me, and where I'd have some sort of a chance. I've never had any here. You can put me in jail now, if you like — they'll feed and clothe me there, anyhow, and I'll be on a level with the rest."

The boy had blurted it all out sullenly and half-chokingly. A world of rebellion and protest against the fate that had always dragged him down was couched in his voice.

Miss Calista drew Dapple to a standstill before her gate.

"I'm not going to send you to jail, Chester. I believe you've told me the truth. Yesterday you wanted me to give you Caleb's place and I refused. Well, I offer it to you now. If you'll come, I'll hire you, and give you as good wages as I gave him."

Ches Maybin looked incredulous.

"Miss Calista, you can't mean it."

"I do mean it, every word. You say you have never had a chance. Well, I am going to give you one — a chance to get on the right road and make a man of yourself. Nobody shall ever know about last night's doings from me, and I'll make it my business to forget them if you deserve it. What do you say?"

Ches lifted his head and looked her squarely in the face.

"I'll come," he said huskily. "It ain't no use to try and thank you, Miss Calista. But I'll live my thanks."

And he did. The good people of Cooperstown held up their hands in horror when they heard that Miss Calista had hired Ches Maybin, and prophesied that the deluded woman would live to repent her rash step. But not all prophecies come true. Miss Calista smiled serenely and kept on her own misguided way. And Ches Maybin proved so efficient and steady that the arrangement was continued, and in due time people outlived their old suspicions and came to regard him as a thoroughly smart and trustworthy young man.

"Miss Calista has made a man of Ches Maybin," said the oracles. "He ought to be very grateful to her."

And he was. But only he and Miss Calista and the peppermint bottle ever knew the precise extent of his gratitude, and they never told.

Becoming a Judge Vindicates her Father

♦♦♦

BY

JUDY

STEED

When Maryka Omatsu gave a reading last fall from her book, *Bittersweet Harvest,* describing her father's life, she looked up and saw that her mother was crying.

"It was the first time I connected with my mother about what the family had been through," says Omatsu, who never shared that feeling with her father: "When he died in 1981 at the age of 80, we were virtual strangers."

The youngest of three children, Maryka was then a criminal lawyer and leading figure in the field of human rights — though she still did not know what her own family had experienced when Japanese Canadians lost their human rights during World War II.

This month [Feb. 1993] she will be sworn in as an Ontario Provincial Court judge, becoming the first Japanese Canadian appointed to the judiciary in Canada. In doing so, she vindicates her father's suffering: "In his last years, my father's pride dictated that he prove that the years of struggle, hardship and repression that he endured in Canada had been worth it; that it hadn't all been in vain," she wrote.

After his death, she was required to go through Denno Omatsu's papers and found, in his wallet, next to his senior citizen's card and Canadian citizenship card, "his worn, World War II government-issued 'enemy alien' identification card, complete with his photo and thumb print."

Prime Minister Mackenzie King imposed the War Measures Act in 1942, following Japan's bombing of Pearl Harbor in December, 1941, and forcibly removed 22 000 Japanese Canadians from their British Columbia homes; during internment, Denno Omatsu was jailed for "travelling without authority"

} BECOMING A JUDGE VINDICATES HER FATHER

from one detention camp to another, looking for work.

But he'd hidden his "shameful secret" from Maryka, who was born in 1948 in Hamilton, where the Omatsu family settled after the war. Her parents never spoke of what they'd endured, didn't want her to know about their losses, hoped to shield her from the shame of internment.

Denno's secrets remained buried until the 1980s, when Maryka became involved in the struggle for redress, travelled the country, met "the community" that had been forever dispersed, heard the war stories, and helped steer the group to victory.

In 1988, 15 days after President Ronald Reagan signed a $1.25 billion compensation deal with Americans of Japanese ancestry, Prime Minister Brian Mulroney — who had long refused to return phone calls from the National Association of Japanese Canadians — signed a $400 million compensation package. It included a $21 000 lump sum payment to 18 000 Japanese Canadians whose property had been confiscated.

Today, speaking at a panel on racism at University of Toronto law school, Omatsu is seated between Rodney Bobiwash, native leader and "Klanbuster," and Bernie Farber, son of a Holocaust survivor and activist with the Canadian Jewish Congress.

Bobiwash and Farber represent groups who have faced genocide, whose plight is perhaps better known; the law students don't seem to grasp that Japanese Canadians really were the victims of racism right here in Canada, less than 45 years ago.

"Japanese Canadians came here 115 years ago," Omatsu says to the law students. "For 71 years we were subject to legal racial discrimination, denied the right to vote, paid half the hourly rate of whites, and until 1949 were barred from professions of law, pharmacy, teaching etc."

Self-effacing, soft-spoken, Omatsu does not tell them how the racism of her Hamilton neighbors and schoolmates made her feel like E.T., the lonely extraterrestrial. Doesn't tell them about the evening her father came to walk her home from a Brownie meeting. As she held his hand, "excitedly describing the secret knots I had just learned, a group of boys yelled at us, 'Chink, Chink.'" She dropped her father's hand "and if I could have I would have crossed the street. We walked home in silence. I still carry with me my childhood feelings of fear and self-hatred. There is no inoculation against this virus."

To the law students, Omatsu explains the legacy of the struggle for redress: "I may agree with David Suzuki that redress may have been too little, too late, but it was better than nothing, and it established a powerful precedent protecting minority rights."

After the seminar, we stop for tea and scones at the Intercontinental Hotel on Bloor, where she notes, with pride, that the beautifully designed interiors are the work of Bruce Kuwabara, a Japanese Canadian architect.

How does she feel, about to leave behind her activist life, to step into judge's robes, no longer allowed to speak out publicly on issues? She smiles and seems unintimidated by the prospect. She's accustomed to the judging part: She's been the chair of the Ontario Human Rights Tribunal since 1991, and from 1979 to 1985 directed the federal government's human rights office in Ontario.

Omatsu's interests are so eclectic that she hops from topic to topic with the ease of an energetic grasshopper. Not that she's flighty: far from it. Everything she touches, in her gentle manner, reverberates with meaning.

Sipping tea, eating scones, she talks about how she juggled a diverse criminal law practice with human rights and environmental activism, and a family. She is married to University of Toronto philosophy professor Frank Cunningham, whose son by a previous marriage she helped raise.

She worked as legal counsel for, among other groups, the Ontario Metis and Aboriginal Association, and the Canadian Paperworkers Union, acting on environmental issues — a long-term interest. Of the major "cases" she's played a lead role in, she counts fighting for the Grassy Narrows Indian band in the '70s as one of the most important.

"We brought experts from Japan who explained that mercury poisoning, or minimata disease, was causing the problems suffered by the people of Grassy Narrows." Although the damage was done to the people and the environment, pulp mills are no longer allowed to emit mercury into the water.

She talks about the bond she feels with Native Canadians, with Jewish Canadians. About how Japanese Canadians had a lot to learn from the Jewish community, whose great gift to other racial groups and minorities, she says, was its persistent exposure of the trauma of racism. Listening to the children of holocaust survivors talk about how their parents passed on the unspoken pain, she understood: "That applied to us, too, but we hadn't thought of ourselves as traumatized. We were a classic textbook case, but we didn't know it."

"Secrets: dear father, if only you hadn't weighed yourself down with so many secrets," she wrote. "Never once did we speak of your war time experiences. Unbelievably, you let me learn about the most central event in your life from my Grade 12 history textbook, which reduced your incarceration, property confiscation and degradation to four lines."

There was much to understand, she said, about how victims of racism internalize negative messages and hate themselves as they are hated by the dominant white society — their crime being merely that they are different.

The denial of her parents' generation, she said, was motivated by fear, "fear of a huge racist backlash that would rise again." When Omatsu became involved in redress, "my mother told me not to demonstrate 'or the RCMP will take you away.'"

Her mother and sister initially claimed "that the Mackenzie King government's actions had been for our own good. My sister, like many others, still maintains that the community's internment and uprooting were on balance beneficial because the experience destroyed once and for all times the Japantown ghetto and forced us to assimilate."

Perhaps the only good thing Omatsu associates with the postwar period, when she was a child, was that her mother Setsuko became "liberated." Setsuko had been a submissive wife back home in Vancouver, where Denno owned a restaurant. When they landed in Hamilton, Denno was too old to establish another career and would soon retire. Setsuko went out to work.

"She'd had long hair in a bun, and she cut her hair," Maryka said. "She bought suits, and she went to her job every day. It wasn't a great job (in a dry cleaning plant), but it changed the power balance in the house."

Like her mother, Maryka always worked. At 11, she had a job as a page in the local library, where she quickly read her way through all the books in the children's section. An outstanding student, she dreamed of working for the Canadian Broadcasting Corporation. In high school she wrote for the yearbook; at the University of Toronto, she wrote for *The Varsity,* was a stringer for *The Star,* and belonged to a crowd that included Premier Bob Rae, Jeff Rose, now a key adviser to Rae, and Michael Ignatieff, now a writer in England.

After graduation in 1971, she took a year off and travelled around the world by herself; she wrote a novel while living on a houseboat in Kashmir: "It was exquisite. The water was still, covered with open water lilies, with the Himalayas in the background."

Her travels shook her out of her artistic fantasies. "I was shocked by the magnitude of the problems, the poverty in India, the suffering in Vietnam." She'd visited Vietnam while the war was on.

Back at home, after a masters in sociology (thesis on women in the Canadian labor force), she ended up in law at Osgoode Hall. She articled with Charles Roach, a prominent African-Canadian lawyer and anti-racist activist, with whom she practised for a few years. The culmination of her work in that era was the famous Jamaican mothers' case in the late '70s.

"People could come to Canada to work," she says, "but they weren't allowed to stay." The case revolved around seven women imported as nannies; they weren't allowed into Canada if they had children, so they said they didn't have children.

When they'd established themselves here, "they tried to bring their kids in. That was grounds for deportation, because they'd lied on their immigration forms. We lost legally, we fought it all the way to the Supreme Court, but we won politically. The law changed."

Did she like criminal law? "I found it really hard. Criminals are always picked up in the middle of the night, you're going to jails at all hours, you work long hours — I had a legal aid practice — and I was getting worn out."

The other difficult aspect had to do with discrimination: "The top women in criminal law, like Marlys Edwardh, all talk about the glass ceiling. The better you are, the more serious cases you do, but the more serious cases don't want women lawyers. Most women leave criminal law after five years."

And now Maryka Omatsu disappears from public view. She is not allowed, as a judge, to sit on boards that raise money from the public. If a cabinet minister calls her, she knows what to say: "Maybe you don't know it, but you're not allowed to call a judge."

She knows that in 1981 when Tom Berger, then a judge in the British Columbia Supreme Court, spoke out on aboriginal issues, he was criticized by Bora Laskin, chief justice of the Supreme Court of Canada.

Berger resigned.

Omatsu figures she can deal with it. She quotes from an old judge's saying: "I can't promise you an empty mind, just a fair one."

Slocan Diary

◆◆◆

BY

KAORU

IKEDA

The Days Before Removal

Our men who had been working in factories, lumber camps and other places were all dismissed and exiled from Vancouver. They were sent to work in road camps in Ontario, Alberta and the interior of British Columbia. They were separated from women, children and older people unable to work, who would be placed in barracks in Hastings Park. Since those who live in Vancouver were also being gradually sent away to road camps there was so much confusion in the Powell Street area that you would have thought it was the scene of a fire. At the beginning of March, K.-san was sent to a place called Hope. A curfew was imposed from March 2nd, forbidding any Japanese to be outside between seven in the evening and seven at morning. Against these and other pressures we have not a single weapon to defend ourselves. As enemy aliens we could not claim the protection of the law and even if we were kicked around we could do nothing but grit our teeth, swallow our tears, and obey orders.

In the middle of this turmoil, Mr. K.T., who had been sick since the year before, passed away. I heard that, unfortunately, a sudden decline in the state of his health came in the middle of the night. A doctor was phoned but the police would not permit a visit nor would they allow anyone to run over and fetch Mrs. H., Mr. K.T.'s sister who lived one block over. By dawn he was dead. The thing his wife found the most regrettable was that he could not receive a doctor's care up to the end.

My son-in-law Tokunaga, even after taking ill last year, took on all the administrative responsibilities for a lumber company, which allowed him little time for rest; even when he got sick with the flu he could not take time off. On top of this the tension he felt when negotiating with hostile whites and the worry of having our future livelihood snatched away, had tired him body and soul. Chisato was gravely concerned as Tokunaga became painfully haggard and drawn.

My son Arimoto who had left Ocean Falls and had joined us in Vancouver was living with us as he had not been told to report to Hastings Park. But he was finally sent away to a road camp, a place called Taft, on April 13th. Chisato and the rest of us were greatly worried that Tokunaga would also be sent to a camp but fortunately the doctor's diagnosis was that while he was not sick, he was definitely not eligible to work. To my tremendous relief he was given permission to relocate with the rest of our family. Wealthier people were given certain choices as to where they could live on their own resources. They were allowed to take all of their household goods with them, but the burden was a heavy one because they had to pay all moving and living costs themselves; naturally most people cannot afford it.

Each of the church denominations was co-operating with the Security Commission on behalf of its followers, to rent a plot of land where they could be moved to. It seemed natural for the Japanese pastors who worked hard for their compatriots and followers, but I could not help feeling grateful and happy to the point of shedding tears to see *hakujin* missionaries work devotedly to help the Japanese. The necessary arrangements for places to live and fuel were to be handled by the Commission. The only thing we would have to pay ourselves would be the cost of food. This way the expense would be quite low. My family were not really members of a church but when we heard that non-members could go with them we applied to the Anglican church.

Catholics were going to Greenwood, Anglicans to Slocan, the United Church people to Kaslo and the Buddhists to Sandon. All could have been described as "ghost towns." Decades earlier they had been prosperous mining towns, but the mines had been abandoned and the buildings, large and small, were decrepit and deserted. They were to be renovated under the auspices of the Commission and turned into homes. I couldn't imagine what sort of places they were. In May we were finally leaving for these camps.

An agent took charge of the house and all our business matters were to be entrusted to him. We were told that the house rent was to be kept by the government and we were not to receive it. We sold off most of the furniture but since the buyer took advantage of our situation we got very little for it. I was mortified, but there was no point in complaining when almost every Japanese was in the same position. All our possessions, house, land, and store, were listed and handed over to the government Custodian. All we could do was to gnash our teeth, hold back our tears, and head for the interior.

Departure
On Wednesday, June 3rd, one hundred and forty people, including us, were assembled and sent to Slocan, British Columbia. Two trains had gone at the end of May and this was the third one.

The day before departure I cut as many flowers from the garden as I could and went to visit the family graves. I offered the flowers to my dead husband, my son Ken, and my older brother, and bid farewell to them all. Not knowing when I would be able to come back to visit their graves, I could not help but weep. When my husband passed away, although I knew he had a full life, my sadness was profound. However, now that we faced these recent disturbances, I began to feel that his death was timely. With his various connections to Japan, if he had been alive, he would have been put in jail. Even so it was much better to have died before being exiled to a godforsaken "ghost town" and forced into a miserable life in his old age. He was a happy man to die surrounded by his family and friends.

We took rice to the T.'s house, cooked it and made many rice balls to stuff in our lunch boxes. We also packed our bags with bread, butter, roast chicken, canned goods, and fruits to take with us on the long trip. This was enough food for about three or four days. Soon my daughter's family came to join us for a farewell dinner that Mr. and Mrs. T. kindly prepared for us. They made *mame* and rice and wished us good health. Although Mrs. T.'s homecooking was always delicious, that day the miso soup and the boiled food was especially tasty. As the excellence of the food touched me, I wondered if the day would come when we could again share a meal. My heart was heavy.

We said goodbye, got into the car and left for the CPR station, arriving at six o'clock. Friends and acquaintances were there to see us off. The platform was noisy and crowded with so many people saying goodbye. At six-thirty we boarded the train. The Mounted Police examined each of us and checked our names off against a registry. Every passenger was handed a dollar to pay for a meal on the train. I again tried hard to control my tears and be silent while experiencing such humiliation. I kept repeating to myself, "We are no longer Canadian citizens, we are Japanese taken prisoner in an enemy country; behind us stands the national dignity of great Japan. One day we will again have our dignity. One day…"

At seven o'clock the train slowly pulled out. I felt hollow as I was being driven away from the place where I had lived for nearly thirty years. At rail crossings here and there, crowds of Japanese gathered to wave and call out their good wishes. On a bluff near the outskirts of town the T. family were sadly waving goodbye. All of us were moved to wave handkerchiefs out the windows, until we were lost from view.

Civil rights demonstrators protesting police tactics during anti-segregation demonstrations, Annapolis, MD, 1964

That Summer I Left Childhood Was White

♦♦♦

BY AUDRE LORDE

The first time I went to Washington, D.C. was on the edge of the summer when I was supposed to stop being a child. At least that's what they said to us all at graduation from the eighth grade. My sister Phyllis graduated at the same time from high school. I don't know what she was supposed to stop being. But as graduation presents for us both, the whole family took a Fourth of July trip to Washington, D.C., the fabled and famous capital of our country.

It was the first time I'd ever been on a railroad train during the day. When I was little, and we used to go to the Connecticut shore, we always went at night on the milk train, because it was cheaper.

Preparations were in the air around our house before school was even over. We packed for a week. There were two very large suitcases that my father carried, and a box filled with food. In fact, my first trip to Washington was a mobile feast: I started eating as soon as we were comfortably ensconced in our seats, and did not stop until somewhere after Philadelphia. I remember it was Philadelphia because I was disappointed not to have passed by the Liberty Bell.

My mother had roasted two chickens and cut them up into dainty bite-size pieces. She packed slices of brown bread and butter and green pepper and carrot sticks. There were little violently yellow iced cakes with scalloped edges called "marigolds," that came from Cushman's Bakery. There was a spice bun and rock-cakes from Newton's, the West Indian bakery across Lenox Avenue from St. Mark's School, and iced tea in a wrapped mayonnaise jar. There were sweet pickles for us and dill pickles for my father, and peaches with the fuzz still on them, individually wrapped to keep them from bruising. And, for neatness, there were piles of napkins and a little tin box with a washcloth dampened with rosewater and glycerine for wiping sticky mouths.

I wanted to eat in the dining car because I had read all about them, but my mother reminded me for the umpteenth time that dining car food always cost too much money and besides, you never could tell whose hands had been playing all over that food, nor where those same hands had been just before. My mother never mentioned that Black people were not allowed into railroad dining cars headed south in 1947. As usual, whatever my mother did not like and could not change, she ignored. Perhaps it would go away, deprived of her attention.

I learned later that Phyllis's high school senior class trip had been to Washington, but the nuns had given her back her deposit in private, explaining to her that the class, all of whom were white, except Phyllis, would be staying in a hotel where Phyllis "would not be happy," meaning, Daddy explained to her, also in private, that they did not rent rooms to Negroes. "We will take you to Washington, ourselves," my father had avowed, "and not just for an overnight in some measly fleabag hotel."

American racism was a new and crushing reality that my parents had to deal with every day of their lives once they came to this country. They handled it as a private woe. My mother and father believed that they could best protect their children from the realities of race in america and the fact of american racism by never giving them name, much less discussing their nature. We were told we must never trust white people, but *why* was never explained, nor the nature of their ill will. Like so many other vital pieces of information in my childhood, I was supposed to know without being told. It always seemed like a very strange injunction coming from my mother, who looked so much like one of those people we were never supposed to trust. But something always warned me not to ask my mother why she wasn't white, and why Auntie Lillah and Auntie Etta weren't, even though they were all that same problematic color so different from my father and me, even from my sisters, who were somewhere in-between.

In Washington, D.C. we had one large room with two double beds and an extra cot for me. It was a back-street hotel that belonged to a friend of my father's who was in real estate, and I spent the whole next day after Mass squinting up at the Lincoln Memorial where Marian Anderson had sung after the D.A.R. refused to allow her to sing in their auditorium because she was Black. Or because she was "Colored," my father said as he told us the story. Except that what he probably said was "Negro," because for his times, my father was quite progressive.

I was squinting because I was in that silent agony that character-ized all of my childhood summers, from the time school let out in June to the end of July, brought about by my dilated and vulnerable eyes exposed to the summer brightness.

I viewed Julys through an agonizing corolla of dazzling whiteness and I always hated the Fourth of July, even before I came to realize the travesty such a celebration was for Black people in this country.

My parents did not approve of sunglasses, nor of their expense.

I spent the afternoon squinting up at monuments to freedom and past presidencies and democracy, and wondering why the light and heat were both so much stronger in Washington, D.C. than back home in New York City. Even the pavement on the streets was a shade lighter in color than back home.

Late that Washington afternoon my family and I walked back down Pennsylvania Avenue. We were a proper caravan, mother bright and father brown, the three of us girls step-standards in between. Moved by our historical surroundings and the heat of the early evening, my father decreed yet another treat. He had a great sense of history, a flair of the quietly dramatic and the sense of specialness of an occasion and a trip.

"Shall we stop and have a little something to cool off, Lin?"

Two blocks away from our hotel, the family stopped for a dish of vanilla ice cream at a Breyer's ice cream and soda fountain. Indoors, the soda fountain was dim and fan-cooled, deliciously relieving to my scorched eyes.

Corded and crisp and pinafored, the five of us seated ourselves one by one at the counter. There was I between my mother and father, and my two sisters on the other side of my mother. We settled ourselves along the white mottled marble counter, and when the waitress spoke at first no one understood what she was saying, and so the five of us just sat there.

The waitress moved along the line of us closer to my father and spoke again. "I said I kin give you to take out, but you can't eat here. Sorry." Then she dropped her eyes looking very embarrassed, and suddenly we heard what it was she was saying all at the same time, loud and clear.

Straight-backed and indignant, one by one, my family and I got down from the counter stools and turned around and marched out of the store, quiet and outraged, as if we had never been Black before. No one would answer my emphatic questions with anything other than a guilty silence. "But we hadn't done anything!" This wasn't right or fair! Hadn't I written poems about Bataan and freedom and democracy for all?

My parents wouldn't speak of this injustice, not because they had contributed to it, but because they felt they should have anticipated it and avoided it. This made me even angrier. My fury was not going to be acknowledged by a like fury. Even my two sisters copied my parents' pretense that nothing unusual and anti-american had occurred. I was left to write my angry letter to the president of the united states all by myself, although my father did promise I could type it out on the office typewriter next week, after I showed it to him in my copybook diary.

The waitress was white, and the counter was white, and the ice cream I never ate in Washington, D.C. that summer I left childhood was white, and the white heat and the white pavement and the white stone monuments of my first Washington summer made me sick to my stomach for the whole rest of that trip and it wasn't much of a graduation present after all.

Demonstrators during the Aug. 28, 1963, civil rights march on Washington

Question and Answer

♦ ♦ ♦

BY

LANGSTON

HUGHES

Durban, Birmingham,
Cape Town, Atlanta,
Johannesburg, Watts,
The earth around
Struggling, fighting,
Dying — for what?

A world to gain.

Groping, hoping,
Waiting — for what?

A world to gain.

Dreams kicked asunder,
Why not go under?

There's a world to gain.

But suppose I don't want it,
Why take it?

To remake it.

from

Where
the Rivers
Meet

◆◆◆

BY

DON

SAWYER

The Corral became strike headquarters. The next day the students circulated in and out, filling up the tables with coke cans and discussing their plans. After the initial rush of the walkout, the cold reality of what they had done began to sink in. About a third of the students had not walked out. They told the strikers that Patterson had called them together in the gym and told them that unless the others returned by the end of the week, the strikers would all lose credit for the year and those in grade twelve wouldn't graduate.

"Can he do that?" asked Roger Charlie, who was hoping to go on into a fisheries program in the fall.

"Naw, I don't think so," assured Jacob. "He's just trying to bluff us." But there was a note of uncertainty in his voice, and Nancy could see the worry growing in Roger's eyes.

"What exactly are we striking for?" asked Charlotte Adams. "I mean, don't people always have definite demands when they strike?"

"We should demand that they start an Indian studies program," someone called out.

"What does that mean?" Bobby McKay demanded. "Traditional stuff like hide tanning and basket making? Spirituality? Or things about what it's like to be an Indian now and what skills we need today?"

"I don't know, maybe some of both."

"Why bother? They'll just teach it like all the other courses."

"I think we need a Shuswap language program."

"Well listen, we've gotta learn the stuff we need so we can make it outside, too."

"Like what?'

"Math, English, that kinda thing."

"But it's more important that we learn what we need here in this community. This is where we live."

"All I know," Lita Adams said quietly, "is that if you come out feeling crummy about yourself at the end then it doesn't matter what they've taught."

"Yeah, and you gotta learn how to think, not just copy notes off the board."

"But look," broke in Roger, "we need programs that are going to get us ready to take vocational courses so we can get jobs."

"What about stuff you need for university? I want to be a teacher."

"The most important thing to me," Big John Joseph said slowly, "is to learn what it means to be Indian. I want to feel good about me and being Indian."

There was a long pause. "Geez, this isn't going to be easy," muttered Lita, who had taken out a piece of paper to jot down their demands.

Nancy looked around the circle of discouraged faces. "Why should it be easy? We're all different and have different ideas about what schools should be and what we need. Why should we agree any more than anyone else? We can't hassle all this out now, but one thing is clear. We need to get to the people who can make changes. We need to have a showdown with the school board and the superintendent. We've got to get them here and listen to us. Patterson will never change. We've got to go over his head."

"So how do we do that?"

"Somebody's got to call and ask the superintendent, I guess."

"Well, who's going to call?"

Everyone looked at Nancy. "Why me?" she asked. "I'm not any leader."

"Maybe not, but since you've come back you're different," said Cyril Narcisse. "You can say things we think but are afraid to say. You do it."

So Nancy walked to the pay phone in the corner and dialed the school board office. She asked for Harry Dworkin, the superintendent, and after several minutes, he finally came on the other end.

"Mr. Dworkin," Nancy began, "I am one of the students here in Creighton who walked out. I guess you heard about that?"

"You must be Nancy Antoine," Dworkin replied. "Yes, Mr. Patterson informed me that there was some problem."

"Well, it's more than some problem. It's a lot of problems. And we would like to talk to you about our complaints and see if we can make things better here."

The phone seemed to go dead for a moment as if the superintendent had put his hand over the receiver. She could hear muffled voices but couldn't make out what was being said.

Finally Dworkin came back on the line. "Yes. Well, Miss Antoine, I'm not in the habit of negotiating with students. I'm afraid that you are dealing with the wrong person. My advice is to go back to classes and then take up your complaints with the administration of your school. Mr. Patterson is a reasonable man."

"But Mr. Dworkin, we..."

"I'm sorry, Nancy; that's my final word. I think it is only fair of me to remind you that all of you who continue to stay away from classes are jeopardizing your year. I might also mention that disciplinary action, including dismissal, may be considered for those who do not attend classes," he paused for effect, "and especially those who seem to be encouraging this behavior. You think about it Miss Antoine."

The phone went dead in her hand. She slowly put it back on the hook and returned to the students who were looking expectantly at her.

"Well?" Cyril asked.

Nancy looked around. Time was obviously against them; the school could simply sit back and wait. One by one the students would drift back and the strike would collapse with nothing gained and people feeling even more powerless than before.

"Well what?" Nancy asked, her eyes flashing. "We didn't expect them to give in with the first shove, did we? Now it's time to organize. We've got to pressure those guys to meet with us."

"How?"

How? thought Nancy. "First of all we've got to get the press here. T.V. and radio and newspapers. Then we have to get the parents, the community on our side. We have to explain why we've walked out. We'll make so much noise they can't ignore us."

People smiled and cheered with relief. Finally there was a focus. There was something to do.

Then Nancy noticed Lonnie Thomas walk in through the cafe door. He wore shiny new cowboy boots, a satiny western shirt with pearl buttons, and a beige sport coat. His jeans were pressed in a sharp crease down the front of each leg. Lonnie was from the reserve and worked as the home-school coordinator. He was supposed to help students who had problems with the school but usually he drank coffee in the staff room and only talked with students to warn them about attendance or behavior. What does he want? Nancy wondered.

He walked up to the circle of students and smiled broadly. His wispy mustache hung down over his gleaming teeth.

"Hey guys, what's happening?" he asked brightly. There was a long silence as people eyed him suspiciously.

He cleared his throat nervously. "Hey, I heard there was a little trouble, so I hurried right down here to see if I could, you know, help work things out."

"Whose side are you on?" Cyril asked, punching a straw through a styrofoam cup. "Did Patterson send you?"

"Hey, we're all on the same side here, right? We all want you guys to get a good education. I'm just here to help patch up the misunderstandings and get you guys back in school where you belong. Listen, I've been working on this and I think I've got Mr. Patterson ready to meet with you guys if you'll come back. Hey, I can't guarantee anything, but maybe we can even get him to agree to a native studies program. What do you think of that?"

There was no response. Cyril kept poking holes in the cup.

"So what'ya say? Let's forget all this and, ha ha, bury the tomahawk."

Cyril stopped poking the cup. "Hey Lonnie, I got one question for you. Are you an Indian?"

Lonnie laughed a high nervous laugh. His curly black hair glistened. "Hey, what kinda question is that?" His laughter trailed away until he stood silent, awkwardly shifting his weight from foot to foot.

"Go on, Lonnie," Cyril continued. "We've got work to do. And tell Patterson that we'll talk about coming back after we meet with the school board and superintendent. Not before."

Lonnie's smile faded away and his face became sullen and hard. "Who do you guys think you are? Who are you to demand anything? I'm gonna give you a piece of advice and you better listen good. If I were you I'd get back in that school fast. Right now they're still not too mad. They may let you back in. But if you continue this stupid little protest much longer you're in big trouble."

Big John Joseph stood up and began walking toward Lonnie, his thumbs hooked in his belt. "You better go now," he said, moving forward threateningly.

Lonnie took a step back. "OK, OK, have it your own way. Throw away your education. But you haven't seen anything yet. This is gonna get nasty." He gave a dark sneer. "You guys are pathetic. You actually think you can bring the school down?" He stalked away, giving the door a vicious shove on the way out.

There was an uncomfortable silence after he left. "Maybe he's right," someone muttered. "Maybe we're just kidding ourselves."

"Maybe he *is* right," Nancy said quietly. "I don't know what we can do either. But we haven't even tried yet. Do you want to give up now and slink back to school?"

Big John, always so soft-spoken before, stopped walking back to his chair and turned to stare at the group. "Not me," he said. "If I go down this time, I do it fighting. I'm tired of being stepped on and feeling weak. I've got nothing to lose."

"Right," someone called. "Let's get to work." There were shouts of agreement, and Nancy felt the tension break. The excitement and momentum was there again. But for how long? They had to force a showdown quickly.

They broke the community up into sections and teams of students formed to go to each house on the reserve. The few white students agreed to try to explain the situation to the white community.

"Good luck," someone muttered.

Jacob smiled. "Look, we can go back to that school tomorrow and there will be no problem. Or we can go to another school that looks just like this one. You're the guys fighting for your lives. It doesn't really matter if we succeed or not. What's critical is that you get the Indian community behind you. Without that you're done for."

Nancy agreed to contact the media. She called the local television station and asked for the news director.

"A hundred Indian students walked out of Creighton High?" the woman gasped. "Are you sure?"

Nancy laughed. "I'm sitting with about a quarter of them now."

"We'll have a camera crew out there in an hour."

That night Nancy watched herself on television for the first time, explaining the situation to the newswoman. The interview had been conducted in front of the Corral, and behind them the cafe's logo — the bronc rider bouncing above his horse — was painted on the glass. On television it looked like the cowboy was putting his hat on Nancy's head. Next time, she thought, we'll do our interviews somewhere else.

She was sitting at George's with a group of other strikers. "How do you think I did?" Nancy asked anxiously.

"Great," George said, "except your zipper was down on your jeans."

Nancy's eyes grew wide. "It was not! Was it?"

Everybody laughed, and Nancy batted George playfully on the shoulder. "Geez George, that's nasty."

"Is that all the coverage we get?" someone asked. "What's going to happen now?"

The next morning the town was crawling with newspeople who sensed a larger issue. Four newspapers, two radio stations, and three television stations, including the CBC, had reporters combing Creighton for anyone involved in the walkout.

"What should we say?" Mary McKay asked.

"Say anything you want," Nancy answered. "But remember what we want: a meeting with the school board and superintendent. Make it clear we're willing to talk; they're the ones who're refusing to negotiate."

Nancy talked until she was hoarse, and that night the news reports were full of the strike. Even Big John Joseph got on television. He watched himself on George's set and a big grin spread across his face.

"Are you one of the strikers?" a young interviewer was asking John while pushing a microphone in his face.

"Yep," said John.

"And do you want a meeting with the school board?"

"Uh huh."

"And what do you want to say to them?"

"Things need to change," John answered.

"This must be pretty exciting, eh?" asked the woman, a note of exasperation creeping into her voice.

"Yep," said Big John.

"Geez," John said, "I've never seen myself on T.V. before."

"You look great," George said, "but next time you gotta remember to talk."

The next morning, Nancy was at the Corral with a group of students working on plans to picket the school. Shortly after the telephone rang, Leo Gamboni, the Corral's owner, called from behind the counter. "Nancy, it's for you."

Nancy walked over to the counter and Leo handed her the phone. "Listen," he said, "I don't mind you turning my restaurant into an office, but if you're going to start taking calls here, how about having a business line put in over there in the corner?"

Nancy put the phone to her ear. "Hello."

"Miss Antoine, this is Mrs. Loretta Lewis. I am Mr. Dworkin's secretary." She had the same cold, efficient voice Doughnut had. Nancy wondered if they learned that in secretarial school.

"Yes."

"I'm calling to arrange a meeting. Mr. Dworkin will meet with you and a small group of students, no more than five, in the principal's office at Creighton High at 8:30 tonight."

Nancy thought quickly. "Tell Mr. Dworkin that we're pleased he's agreed to meet with us," Nancy paused, "and the time's fine. But we'll only meet if whoever is interested can come." There was a sharp intake of breath on the other end. "And we'll meet at the pow wow grounds next to the band hall."

There was a long silence, then a hand was placed over the receiver. Nancy thought she could hear Dworkin's voice in the background, but she couldn't be sure.

After what seemed like minutes, Mrs. Lewis came back on. Her voice seemed even colder than before. "Very well, Miss Antoine. 8:30 at the, er, pow wow grounds."

Nancy exhaled in relief. "Good. Oh, and we want the school board members there as well."

"Anything else, Miss Antoine?" the secretary asked icily.

"No. That should be fine." She hung up the phone and rushed back to the waiting students. "We've got some new planning to do," she said simply. "The meeting's on for tonight."

By 8:15 Nancy guessed that at least two hundred people had gathered at the pow wow grounds. The student teams had fanned out through the community, revisiting each house on the reserve to tell them about the meeting. Now more and more people crowded into the clearing.

Roger Charlie had brought a truckload of firewood that afternoon and they had built a fire in the fire pit. The blaze crackled noisily in the centre of the circle.

Nancy watched the people. There were old women, bandannas tied around their heads. Young mothers with two or three kids in tow. Older children running and laughing, dogs barking at their heels. Men in jean jackets and worn cowboy boots. She saw George talking with a group of workers from the mill. Then, walking down the dirt road in front of the band hall, she saw a big man walking with long strides. Beside him, dwarfed by his size, was an older woman who walked with surprising spring and determination. It was her father and Mrs. Schmidt.

Nancy ran to them and hugged them both. "I didn't think you'd come," she said, smiling with delight.

"You kidding?" asked Mrs. Schmidt. "I wouldn't miss this for the world. Besides, teachers always like to see their students show what they've learned."

Just then two shiny new cars bounced along the road and pulled to a stop. Several women in high heel shoes and dresses climbed out carefully and were helped by men in suits and ties. Nancy's father reached over and squeezed her hand. "Good luck," he said.

Most of the people from the cars stood whispering together and eyeing the crowd uneasily. One man opened one of the trunks and pulled out a large flip chart. Nancy approached him.

"Mr. Dworkin?"

The man stood up and smiled. "That's right." He extended his hand. As Nancy shook it she examined the superintendent. He was younger than she'd expected, probably not more than forty-five, and he was, she conceded, quite handsome. "You must be Nancy Antoine. You have quite a crowd here. Shall we get started?"

Dworkin carried his flip chart through the crowd and set it up near the fire. Then he turned to Nancy. "Perhaps you'd like to begin?"

Nancy hesitated. She looked at the hundreds of faces in front of her and wondered if she could tell them what she felt in her heart. She began to feel some of the old fear again.

"As you know," she began, and the crowd slowly quieted. "As you know, most of the Indian students and some of the white students walked out of school three days ago. We have asked Mr. Dworkin, the superintendent, and the school board members," she waved at the men and women who had come in the cars and were now sitting uncomfortably on the grass, "to come and discuss our complaints. We hope everyone will have a chance to say what they think and," she faltered, "and maybe together we can find some solutions."

Nancy began to get over her nervousness. She glanced over at Mrs. Schmidt, who smiled encouragement. "I recently studied with one of our elders, and I came to realize that there are things we can learn, we need to learn, that your schools can't teach us." She looked at Dworkin. "There are many things to learn from the white world. We need that knowledge and those skills too. But there are also many important things that only our people, our elders, can teach us. We need to learn from each other.

"I don't have a degree from a university, but I have spent a long time in school. I know that there are ways to teach and learn that your teachers are not taught. There are goals and needs that you do not understand. There is information we need and skills that are not taught in your classrooms. And there are places to learn outside of your schools. We have to find and use them."

She stood silent and looked out at the crowd. There was a murmur that ran through the people, but she couldn't interpret it. Had she been too general? How could she explain what she felt?

Behind her Dworkin cleared his throat. "Yes, well, thank you, Nancy, and I want to assure you that we are always open to suggestions and ideas for improving our schools. However, I want to explain to you people the situation we're in. You see, our hands are really tied." His voice was reasonable, almost apologetic. He flipped to the first sheet on his chart. "You people behind me might want to shift over here so you can see this."

"Now here are the requirements for graduation that the provincial government supplies. As you can see, you must take at least eight at this level, including three here." Dworkin had a pointer he pulled out of his pocket and extended like a telescope. He stood lecturing, pointing at the graphs and diagrams, and the murmuring died down. Slowly Nancy began to see the blankness on people's faces replace the interest and excitement that had been there before. It was like watching students in a classroom, and her heart sank.

As Dworkin spoke the sun set and the sky began to turn pink. The fire seemed to brighten. Though the evening was warm Nancy shivered. She felt self-conscious in the centre, and she melted into the edges of the crowd, slumping to the grass.

"So as you can see," Dworkin finished, "we really are doing the best we can. There just isn't room for the kind of changes these students are talking about. But we know, too, that they just don't understand the intricacies and complexities of the situation. It takes those of us working in the field years and years of training, and even then," he chuckled slightly, "we don't always understand it all. So I'm sure that we can get together on this and patch up our differences."

Dworkin beamed at the crowd, and they sat silent and confused. Nancy sensed people beginning to turn away, drifting back into the growing night. A few more moments and it would be all lost. But she, too, felt subdued, too weary to speak up and try to counter the charts and numbers, the crisp logic.

Suddenly a big man pushed his way into the circle and stood awkwardly in the firelight. Nancy realized with a start that it was her father.

He started speaking slowly. "When I was a boy..." He stopped, looked around nervously, then started again. "When I was a boy my grandfather told me a story about his father, my great-grandfather, and his vision quest." Nancy looked at the faces of the school board members sitting hear her. Their expressions were mixed: annoyed, confused, patient. One man looked at the woman next to him and almost imperceptibly rolled his eyes.

"One morning while he was on his quest, he had gone out to pick berries. It was early and the mists were rising off the river. A bear, a grizzly bear, came out of the mist like a spirit. It moved slowly toward him, shuffling." Now her father moved into the centre of the circle and the fire played on his long hair and great shoulders. "He was frightened and wanted to run, but he couldn't. Slowly the big bear came toward him, sniffing the air." Her father's voice seemed to take on a growl and he swayed slightly from side to side. His shadow flickered and danced, huge and bear-like in the half-light of the fire.

"She came to him and he could see the white tips of her thick hair, the deep black of her nose, the shiny black stones of her eyes. She sniffed his hand, then gently, licked the berries from the basket he'd left on the ground. Her pink tongue lifted the red berries into her mouth."

Nancy looked at her father in astonishment. He seemed large and remote. His deep voice had become almost a chant. The red fire flickered in his eyes. As he moved around the circle, he faced the school board members, and new looks passed across their faces.

"That bear finished and walked back into the trees, but before she did she looked at my great-grandfather and told him that she would always be there to help him, to guide him, to give him strength when he needed it. And when he went to get his basket, there was a single bear's tooth lying in the bottom."

Nancy felt a shiver run up her back, and the tooth she still wore seemed to tingle against her skin.

"But you see," her father's voice seemed to soften, "I'd forgotten that story. You and your schools tore those stories out of me. You tried to kill the stories and their strength and leave me empty. And it worked too. But now I'm beginning to remember, and I won't forget again. And now some of our children are beginning to look for those stories and for that strength, and it's not in your schools. Your schools leave them feeling empty and weak, just like they did me."

The circle was hushed, and the superintendent seemed to retreat behind his flip chart as Nancy's father, large and dark in the firelight, faced him.

"Mr. Dworkin, you say you can't make room for us, our lives, our history, our stories, our future, in your schools. If that is true, Mr. Dworkin, then we have no choice but to make a school of our own."

A long silence filled the circle. Suddenly Lonnie Thomas leaped to his feet. "Now let's not be hasty," he said, smiling at the school board people. "These good people have come here to talk with us. Let's not scare them off, ha, ha, or threaten them. I mean I'm sure we can fix things up and, uh, our differences I mean...It takes a lot of work to start a..."

"Shut up, Lonnie," someone yelled. Lonnie's smile faded, his voice faltered, and he slipped back into the crowd.

If Dworkin had cast a spell, Nancy's father had woven a counter-spell. Now Dworkin was subdued. Nancy felt the shift. She shook her head. Their own school! Of course! Why hadn't she thought of that?

Now people started to come forward. Lily Charlie began speaking in her clear, heavily accented, gentle voice.

"Oh, yes, we learned a lot in your residential schools," she said. "We learned to steal because we were always hungry. We learned to wash clothes and take orders and to be afraid of whips and straps. We came back to relatives who had never struck a child in their lives and we thought they were weak, and we disobeyed them. We learned that we were ugly. We learned to do only what we were told. We learned to despise our old ways and hate being Indian. For a hundred years, you have been telling us what we've needed to learn. And look where it's gotten us! Maybe it's time for us to decide what we need for ourselves."

Darrel Williams stood up. "And it's not so different now. I went

through all your charts and I graduated. But what did I have? A lot of useless facts and no knowledge of who I was. I went to trade school, but I spent so much time trying to get myself straightened out I flunked. And now I don't fit in here either."

The rush became a torrent. People shouted out from all over the circle. "My daughter comes home from school and cries because she thinks she's stupid," someone yelled.

A mother, holding a sleeping baby on her shoulder, stood up behind Dworkin. "Your schools are not like our homes, Mr. Dworkin. We consider our children people, free to explore on their own. We don't force our kids to learn at a certain time, in a certain way. They show us what they've learned when they're ready to, not when we demand it. Our children are not used to having one person control their world. They're not used to being told what to do and how long they have to do it."

The clamor grew. Speaker after speaker talked about their anger, their frustration. Old people spoke about residential schools. "Why, do you know," one woman said, "we were even told how many squares of toilet paper we could use! Two." She raised two fingers for emphasis.

"The first time I met the brother," another related, "he pulled out a big black leather strap and told me, 'If I ever hear that you're speaking Shuswap, you'll get this over your hand.' And he did it too."

Another began. "When I went to school the brother asked me my name. 'Mouse' I said, for that was my Indian name. And that brother got the whole class to laugh at me, to ridicule me. Finally he gave me a new name. Norman."

There were younger people who talked about their school experiences. "In twelve years, I was so scared of teachers I don't think I asked a single question," one woman said quietly.

"You know I can hardly write now, and all that stuff I learned doesn't seem to have anything to do with my life," added another.

Abel Charlie spoke. "All that algebra and all, that's only any good, maybe, if you leave here. What if you want to stay? When do we learn what we can do to make this a better place?"

Dozens of people spoke and hours sped by. Nancy looked at her watch. By the firelight she could just see the hands. It was after midnight. Finally, the torrent stopped. The last speaker sat down, and the fire began to burn low in the pit. Nancy looked around. Hardly a person had left.

There was a pause that stretched for a minute, two. Then

Dworkin, who had remained the whole time, wheeling to face speaker after speaker, ran his hands through his hair. He cleared his throat.

"I, I really don't know what to say," he stammered. Then he paused again. "One thing seems perfectly clear, however. We have been running our schools our way, and you people have been left out." He cleared his throat again. "We have made many mistakes for you. Maybe it's time that you had a chance to make your own for a change. If you wish to start your own school, I will not oppose it."

There was a hush, then someone clapped, and then the applause tore through the crowd. People began to cheer and laugh, and Dworkin stood smiling wearily in the middle.

Somebody was shaking Nancy's hand. "It looks like we've won!"

Nancy nodded and made her way through the crowd. She found her father sitting quietly outside the circle. She rushed to him and hugged him. He started to put his big arms around her, hesitated, then clasped her close to him, laughing as she buried her face in his shoulder and wrapped her slender arms around his neck.

"I'm so proud of you, Pop," Nancy whispered. "You were great. Thanks."

Her father unlocked his arms and looked shyly at his daughter. "I felt kind of foolish at first, but then it was as if someone else was telling the story. It seemed to come from way deep inside."

Nancy hugged him again and handed him something she'd been squeezing in her fist. It was the bear's tooth hanging from the thong. It glinted red in the firelight.

"Thanks for lending this to me, Pop," she said softly, "but it belongs to you."

She took it from his hand and slipped it over his head.

Speaking the Language of the Soul

♦♦♦

BY

MARY LOU

FOX

Even after 40 years, I can still hear their voices. "Remember your connection to the land. It's the land that makes us look, act and talk the way we do," my grandfather says in Anishnabe.

"You'll never amount to anything on the reserve," my white teachers retort in English.

My mother was a school teacher on Birch Island Reserve near Manitoulin Island and was caught between the two. She spoke fluent Anishnabe, but it was forbidden in the classroom. In any case, she believed, as her own teachers had told her, that the language would do me no good. The life she wanted for me was written in English.

In those days, I'd see a group of Anishnabe kids at a bus stop, and wouldn't know what language to use. Anishnabe was the sound of shame. My friends went to residential school where you could be punished for speaking the language. I had my own private hell, as the only Indian in an all-white high school in Espanola. The only place I could feel good about myself was back on the Wikwemikong Reserve on Manitoulin with my grandfather. My love for him was tinged with guilt.

Many of my friends, and my mother's friends, grew to adulthood with the Anishnabe skills of a five-year-old. This was a Canada where Indians couldn't vote, or meet publicly. Our ceremonies were held in secret, far back in the bush. Yet we tried desperately to fit in. We learned the strange sounds and expressions of English so well that they almost seemed our own. Perhaps we really believed, like my mother, that we could change who we were.

Waubegineese, my grandfather, knew better. So did the elders. "If you take away the language, you take a part of our soul," they tried to tell us. "Without the language we will cease to exist as a people." Yet such was the strength of the schools and the church that for a time people stopped listening to the elders. English came to be used more and more on the street, in people's homes. A whole way of life that was tied to the Anishnabe language — the songs, the dances, the healing circles, our systems of knowledge and values — began to falter and fade. We gained nothing that could replace it.

My friends and I grew up. We started families. We began to notice that the children weren't speaking Anishnabe among themselves. Many parents who still spoke the language weren't teaching it to their children. Perhaps they hadn't talked to their grandfathers. Perhaps they hadn't even known their grandfathers, since the church and the state had been working for a century to break down the extended family. Whatever the cause, the Anishnabe language in our community was being steadily eroded.

When we started, with difficulty, to talk about it, we learned that this was going on nation-wide. All but three of our 53 native languages were in danger of disappearing. We heard horror stories that were similar from coast to coast: lack of identity, lack of self-respect, no one who shared with us what we have as a people. We heard for the first time of Huron — once the great language of the fur trade, now dead for 100 years. It seemed incredible that so many of our languages had survived this onslaught. We began to suspect that, far from weak, they were stronger than anyone had dreamed.

At home we started a language nest for pre-schoolers. We developed an Anishnabe immersion program for the primary grades. We formed an elders council. Across Canada, other communities did the same. We started to work together at the regional and national level. Through the Assembly of First Nations, the first new national gatherings on language were held. People paid their own way: hundreds came. There is still a tremendous sense of excitement as we discover how much we have in common and how, independently, we have been working in parallel. Twenty years ago I would never have thought something like this could happen. And it's growing every day.

Of course, we still face the same problems. Ironically, in a country where language rights are so often in the news, we have to sell raffle tickets and hold auctions to help our languages survive. It's said that languages that represent thousands of years of first nations civilization and which are unique to this country are neither recognized in the Constitution nor protected under federal law. The federal government does not even have a policy on aboriginal languages.

We would like our vision to be shared. We see a country where aboriginal languages receive equal protection and promotion to that of official languages. Where education in the language of the community is recognized as a right, and sufficient funds are made available for it to succeed. Where our children grow up fluent in two languages and confident in two cultures. Where first nations people with fluency in aboriginal languages are employed as instructors, translators, publishers, journalists and broadcasters, and our cultures are accepted as a vital part of the Canadian mosaic.

We are prepared to do whatever is necessary to bring this about. Our elders are teaching us to look inside us to find the strength and purity we need. For the first time that I can remember, our young people are coming back from the cities to consult the elders and learn from them. Problems that we haven't talked about for years are coming out into the open. We have begun a healing process which involves the whole spirit and way of life of first nations. With our languages at the centre.

Thank you, Grandfather. *Meegwetch.*

Statement of Vision Toward the Next 500 Years

from the Gathering of Native Writers, Artists and Wisdom Keepers at Taos, October 14–18, 1992

In memory of more than 500 distinct Native Nations and millions of our relatives who did not survive the European invasions and with respect for those Indigenous Peoples who have survived, we make this statement.

We, the Indigenous Peoples of this red quarter of Mother Earth have survived 500 years of genocide, ethnocide, ecocide, racism, oppression, colonization and christianization. These excesses of western civilization resulted from contempt for Mother Earth, and all our relations; contempt for women, elders, children and Native Peoples; and contempt for a future beyond the present human generation. Despite this, we are here.

Since time immemorial, Native Nations have lived in harmony with this land and in solidarity with our relations. Our continued survival depends on this vital relationship. We perpetuate this harmony for our continued survival and world peace. We carry out our religious duties for the good of all. Endangering us endangers us all.

We call for the immediate halt of the abuse, neglect and destruction of life. We call for immediate strategies and compacts to halt the genocide of Native Peoples throughout the western hemisphere.

We demand an end to all exploitation, desecration and commercialization of Indian spirituality and cultures, our sacred places and the remains of our ancestors. We demand an end to the violations of our right to worship, to the disrespect of our religious and cultural property and to the disregard of our very humanity. Native Peoples over the next 500 years must maintain our status as distinct political and cultural communities. Indian Nations expect the world community to honor and enforce treaties that recognize tribal property and sovereignty. Sovereignty is the inherent right of Indian Nations to govern all action within their own countries based upon traditional systems and laws that arise from the People themselves. Sovereignty includes the right of Native Nations to freely live and develop socially, economically, culturally, spiritually, and politically.

The domestic laws of the non-Native countries of this hemisphere have been used to subjugate Native Peoples. Vindication of our rights must be achieved through fair and appropriate procedures, including international procedures.

Indigenous Nations have the right to secure borders and fulfilled treaties for which we gave up vast territory and wealth. Native Nations have the responsibility to provide a safe and secure environment for their people's economic self-sufficiency, health and well-being. A secure and adequate land base and respect for sovereignty are prerequisites for viable tribal economies.

Indigenous People have the right to educational and social systems that affirm tribal cultures and values that promote physical, spiritual and mental well-being of people and that teach the care and healing of Mother Earth and all Her children.

We envision that in 500 years Indigenous Peoples will be here, protecting and living with Mother Earth in our own lands. We see a future of coming generations of Native People who are healthy in body and spirit, who speak Native languages daily and who are supported by traditional extended families.

We look forward to leadership that encourages the religious and cultural manifestations of our traditions, and the reclamation and continuing use of traditional ceremonies, hairstyles, foods, clothes, music, personal and tribal names, and medicines. Our cultural renewal will assure the perpetuation of natural species that are dying, and perhaps even some of those thought to be extinct.

We celebrate our rich, continuing tradition of artistic excellence. The works produced for tribal functions or within a religious or historical context are the sole cultural property of the Native Peoples. Our strong cultural continuums accord great freedom of nature of our traditional cultures. We envision a future when our artistic gifts are recognized fully for their spiritual transforming power and beauty.

Native Peoples are strengthened by relations among each other at all levels of community life. Commitment, integrity, patience, the ability to build consensus and respect are essential components to the flourishing of culture, friendship, strengthening of economies and the pursuit of a common peaceful world.

All life is dependent upon moral and ethical laws which protect earth, water, animals, plants, and tribal traditions and ceremonies. Humanity has the responsibility to live in accordance with natural laws, in order to perpetuate all living beings for the good of all Creation. We share a bond with all the world's Peoples who understand their relationship and responsibility to all aspects of the Creation. The first of these is to walk through life in respectful and loving ways, caring for all life. We look forward to a future of global friendship and the integrity of diverse cultures.

Agnes Macph
Reformer

◆◆◆

BY

DORIS

PENNINGTON

. . . .

Agnes Macphail, reformer and politician, was elect-
ed to Canada's parliament in 1921, the first woman
ever to be elected in Canada. She served until she
was defeated in 1940. In 1943 she was elected to the
Ontario legislature, lost her seat in 1945, and then
served again from 1948 to 1951, although her health
was failing. The following describes some of the later
years in her career.

Interviewed by Helen Beattie for the *Canadian Home Journal,* Agnes
said that her principles had not changed over the years.

> *If anything, I have become more radical. But on small things I*
> *have changed. I wouldn't make a fuss about the little things any*
> *more. At one time it worried me terribly that Parliament opened*
> *Thursday and settled down to business on Monday. Those days in*
> *between were to me an awful waste. Now I feel that if it is tradi-*
> *tion it's O.K. by me. I'll just amuse myself over the weekend and*
> *wait until the boys are ready. If I had had as much sense at thirty-*
> *one as I have now, I never would have had a breakdown in health.*
> *I realize too that there is nothing so powerful as an idea whose*
> *day has arrived. Timing is the thing! For example, right now we*
> *are passing through an age of social consciousness and many*
> *things are being accomplished now that could not have been done*
> *a generation ago — no matter how hard a person worked. Look at*

family allowances. The time for them had come, and even those who originally fought against them finally voted for them.

She thought women did not go into politics because they thought it unladylike. But government suffered because "we are only governed by half the human family." She defended women, though, stating that they paid such a high price for human life that they had little energy left for politics.

"Of course, women are no good to fight for themselves. The very fact they slave so hard for their husbands and children proves that... And they aren't loyal enough to each other. Men have learned loyalty to their fellow-men through the years...When I hear women talking it appalls me. They worry about their children's schools, their friends and their clothes — but they haven't the foresight to try to do anything about the kind of world their children are going to live in."...

The [Ontario] Legislature opened in February, 1949. Agnes continued to fight for the weak and oppressed, for the "ordinary" man and woman. She said: "My interest has been and is still — and will I think remain — the people, and the people least able to look out for themselves. I have never been interested in the powerful and the rich, because I think they get more than their share anyway, so I see no reason why I should bother about them."

She had seen a picture of men sleeping on a floor, without blankets. "Many of those men were veterans of the Second World War... When I saw that picture, and then went and saw again this line-up for the soup kitchens standing in the cold winds, waiting to get in, I suffered. I don't know what they were suffering — but I know I suffered. The Honorable Minister of Labour at Ottawa, (Mr. Mitchell) said this was just 'seasonal unemployment.' Well, Mr. Speaker, that does not make the board any easier to lie on; that does not make them any warmer without a blanket. The difficulty is that we do not think about them as if they were ourselves."

Fighting for higher pensions for teachers, she cited a seventy-five-year-old woman who had taught for forty years. Her salary for the first twenty years ranged from $250 to $900 a year; the second twenty years had started at $1400 a year and remained at that rate. She had contributed to the support of her widowed invalid mother

and had had heavy hospital expenses herself (in connection with her arthritis). She gets the magnificent pension of $57.94 a month.

*By the time she pays her board, which costs $50 a month, she has
$7.94 for the amenities, the luxuries of life. She is a very foolish
woman because she pays out $3.50 monthly in order that her bur-
ial expenses may be covered by insurance. I would not care about
that — I would go out to a show and have a good dinner — eat,
drink and be merry.*

*I would like to quote one sentence from her letter which says:
"I can only pray that when the summons comes, the call will be
clear and swift." Well, that does not make me very happy or proud
of Ontario with its boasted resources and well-fed people.*

She fought for improved prisons for women. "Naturally there is nothing like the number of women prisoners there is of men... [but] what I
want is more attention put on the few women there are."

She fought for the elimination of the means test for pensioners. "I
have known cases where their pride is so great where people really
needed the pensions greatly and yet would not apply for them because
they thought it was accepting charity."

She fought for hospital insurance, civil marriages, bursaries for
normal school students, equal pay for women.

On the latter subject she quoted "two or three of the funniest bits"
of a *Maclean's* magazine article by Charlotte Whitton:

*The best one is the last one, so maybe I had better read it so as to
keep everybody cheered up. It is about men and women stuffing
sausages in a packing plant, and Charlotte says: "And why a man
draws 50 to 77 cents an hour stuffing sausages beside a woman
who draws 40 to 53 cents, only the sausages can tell."...*

*Looking at this question simply from the point of view of jus-
tice...you could not argue against it — there is no argument. It is
just custom, and the dominant group governing society. Sometimes
I think it is a pity the men could not have the children; they could
just have this world all to themselves.... They could have all the
jobs and all the pay and all of everything.*

It should not matter whether one had children to support. "I will think
that argument has some meaning when bachelors are not paid as much
as married men for doing similar work; when a man with five children
gets higher wages for certain work than a man with one. That does not
happen; they are paid for the work, and rightly; why should not women?"

The Honorable C. Daley, Minister of Labour, conceded that it had been a contentious issue for years and that "a lot of work in the industrial sphere could be done better by women than men." Agnes interrupted to ask if they got better pay for it.

Daley said the point had been made that Miss Macphail "got the same pay for work in the Legislature. From my own observations over the years, I would say this is probably wrong. Miss Macphail should probably get a little more than some of the men." Agnes said, "You and I agree on that."

Daley said "girls entered industry but tended to stay a shorter time than men." Agnes asked: "Do not men move around from one job to another?"

She thought it was a "disgrace to men that they [were] not willing that women should get the same pay for doing the same work." But some day we would realize that if work was well done, it would not matter if a person were "a man or a woman, white or black, yellow or brown, he or she [would] be paid for the work he or she [did]."

The fall session of the 1951 Legislature was short. In October an election was called. Agnes was defeated along with all but two members of the CCF Party. Her "brave and glorious adventure" was over.

The last two years, however, had seen two of her greatest victories. The government had passed two bills for which she had fought: one the bill granting equal pay for equal work for men and women in Ontario — the other a bill granting old age pensions at age seventy without a means test and at sixty-five to sixty-nine with an eligibility test. She had helped found the Elizabeth Fry Society to rehabilitate female prisoners....

On Thursday, February 11, 1954, Agnes Macphail had a heart attack. She died in Toronto's Wellesley Hospital two days later, a few weeks short of her sixty-fourth birthday.

Tributes poured in from people in every walk of life. Old friend and political leader E.G. Jolliffe said that while she had returned to Grey County, everyone knew that now "she belonged to all Canada....

"When we stood beside her grave, the swift driving snow was lashing the Grey County hills all around us and so fierce was the wind that the voice of the minister could scarcely be heard. Agnes had left the storm-troubled world in which she had done so much for peace and justice...."

"No one in the public life of Canada ever had a wider circle of real friends, and her acquaintances were legion. She had friends in prison and friends in high office, friends who were pacifists and friends who were professional soldiers. To her they were all human, all people, and she loved them for their mistakes and their weaknesses as much as for their admirable qualities."

The March, 1954, issue of the *Pathfinder* (published by men confined to federal prison at Prince Albert, Saskatchewan) wrote a column headed, "We Lost a Friend." It said in part, "On February 14 [*sic*], 1954, this woman whose life was the very embodiment of right and service and enlightened purpose passed to her reward. The man behind the bars lost his greatest benefactor and gained a patron saint.

"May you rest in peace, Agnes Macphail."

Ascendancy

♦♦♦

BY

SHEILA

DEANE

It's my duty day at my daughter's preschool
And I'm in the creative room vacuuming;
Stray bits of playdough, collage, cookie crumbs, sand
(Please *do not* vacuum up large amounts of sand)
Are not being sucked up very well and I begin to think
That maybe I should change the bag when I hear chanting
From the next room "No girls allowed, No girls allowed"
And my own daughter weeping, so I strike off the vacuum,
Run into the room feeling that I'm intruding where the teacher
Should be, but, damn it, she's moving too slow, not moving at all
And it's my daughter weeping, who I take by the hand and pull
To my lap, "What is it?" (I know, I know), "No girls allowed,"
And over her sobbing face I am glaring at the three confident boys
On the climber who know they will win this one
And any one they please because you just can't climb
On a climber if there's a gang ready to push you off
And you have no answer to their taunts, being raised
In kindness and encouragement, no answer to "No girls."

Why *no*? Why no *girls*? It doesn't make sense to her —
She is lost in her grief and its mystery;
I say soothing, hopeless things:
"They don't know that girls can be fun, they just don't know"
(They know what kind of fun they want), I feel the teacher's
Eyes on me, she's smiling, happy that I'm going to sort it out
In a good-girl, good duty-mom way. I say: "You want to play
With them, honey, and that's too bad, but choose something else.
There's Thida, play with her now, sweetheart,
She likes to play with you," and it's over.
Her tears are dried (the wet glaze of her hurt stays in her eyes)
And she runs off to push a baby buggy with the others, round
And round the menacing climber they go, like a crowd of tired
Friday night shoppers, stiff and desperate, under the ladders
Of the boy's boisterous play. The teacher nods, and I turn away

Furious. Hypocrite, good-girl, polite
Coward that I am to my own love's cause,
When I should have simply thrown the climber down in a heap,
Bar by bar, until it was just so much glossy rubble against
The bright shag carpet of the activity room. I know
It would be right, it would be the right message, the heart-sized
Message she should have from me, that I am strong enough,
And love her enough to undo the "No girls allowed" in her life,
That there will be no more "No girls allowed" in her time,
No fortress she and I can't storm, no world we can't open
Enter and possess. Daughter, I go with you in this —
Against harassment, injustice, tyranny, amidst whatever
Toys we find it, no matter what the age of those who shout it;
Daughter you fight for me and I for you in this.
And we cannot refuse to fight.

But I don't really make a scene. I even go back to my
Vacuuming. The image of my dewy-eyed daughter clear
In my mind as she shuffles around the room tensely cradling
A filthy, floppy doll. I don't cry out, break things, weep;
I only wonder under what rumbling mountain I myself
Learned how to crouch and cling this way.

left: Nellie McClung, Manitoba-born writer, reformer, and suffragist; right: Thérèse Casgrain, former senator, reformer, and leader for women's suffrage in Québec

Women Pay More for Identical Items

♦♦♦

BY

CHARLOTTE
PARSONS

At first, Eileen McNamara thought she had made a mistake — it simply didn't seem possible.

The unisex Levis shirt dangling innocently in the women's wear department of Eaton's Ottawa store was identical to the one she had seen just minutes before, downstairs in men's wear. Same colour, same style, same label.

But it was priced at $12 more.

"We were shocked," Ms. McNamara said. "It was cheaper in the men's than in the women's and it was exactly the same. It was $47 in men's and $59 in women's and there was nothing different about it at all."

Ms. McNamara had just discovered a nearly universal fact of economic life: Women pay more.

And not just for clothes. Women are frequently charged more than men for a whole range of goods and services, from dry cleaning to haircuts to cars.

"I call it a gender tax, and it's my hunch that it is more pervasive than anyone has guessed," said Anita Jones-Lee, author of the self-help book *Women and Money.*

One reason for the disparity, she said, is the simple fact that many women are willing to pay the inflated prices.

This is particularly true in the fashion industry, says Montreal designer Richard Doucet, who has been designing and selling clothes for 11 years.

"They make it more expensive because they know if a woman likes it, she'll buy it," he said.

"And there's no other reason, when you're talking about clothes that are similar for both sexes, because men's clothing uses more material, so it should be more expensive."

The higher cost of women's garments doesn't end at the boutique cash register.

Alterations, traditionally offered to men free of charge, stitch extra dollars onto the price of women's wear. In 1991 The Bay stopped providing men with free alterations in response to cries of sex discrimination. But women still pay more than men: $5 for a plain hem ($4.50 for men) and $6.75 for a cuffed hem ($5 for men).

And spaghetti sauce splattered on that newly purchased blouse will cost more to remove than it would from a man's shirt. A survey of 10 Metro Toronto cleaners revealed an average price difference of 63 per cent between a woman's cotton blouse and a man's cotton shirt.

"It doesn't seem right to charge more for a blouse, because they're both cotton," said Anthony Chow, owner of Martinizing Dry Cleaners and Launderers. "But it takes more time. They are smaller and the standard machine doesn't fit them very well."

But boys' shirts and very large men's shirts are also the wrong size for the machine. Mr. Chow said he has to "make a compromise" for such cases by not charging them the higher rate that women pay.

Hair salons also take a larger cut of their profits from female clients. A survey of 10 Metro Toronto stylists found women paying an average of 50 per cent more than men for a shampoo and cut.

And the difference has nothing to do with hair length. Two of the salons surveyed even divided women into long- and short-haired categories, yet still charged short-haired women more than men.

"It's brutal. It's like pay equity — it's got to turn around," said 26-year-old Fran Warburton after a shampoo and trim at a downtown Toronto salon.

"It was $32 for women and $24 for men regardless of length, and I have short hair."

Salon owners protest that women are fussier and hairier. But Patrick Jubé says that's pure nonsense.

"They do it because they can get away with murder," said Mr. Jubé, a stylist for 20 years and founder of the André-Pierre Salon chain in Toronto. "I mean, would you pay more for a dentist because you are a woman? Because it's the same thing. I cut hair for women. I cut hair for men. It doesn't make any difference to me."

Mr. Jubé said the two-tier price system is rooted in tradition.

Decades ago it was the norm for women to spend hours in the

salon while their locks were sculpted into beehives or bouffants. Meanwhile, down at the local barber shop, their male counterparts could be snipped, buzzed and sent on their way with a minimum of fuss.

"Twenty or 30 years ago there were barbers who cut men's hair who weren't as qualified as women's hairdressers," said Mr. Jubé, who sets his prices according to hair length alone.

"So the hairdressers could get away with charging more. But today they are all equally qualified to do both. All you're buying is time. But why drop your prices [for women] 25 or 30 per cent if you don't have to do it?"

But most price discrimination is more difficult to detect. The sight of a female face can swell the price of negotiable goods and services. It is most often where bargaining is involved that women get the short end of the financial stick.

The Montreal-based magazine *Elle Québec* sent a male and a female reporter out to bargain for the same items. They haggled their way through a list of nine different goods and services, from landscaping to computers to burglar alarms.

When the "final offers" were tallied at the end, the female reporter's total was 92 per cent higher than her co-worker's.

"In terms of discretionary bargaining power it's the impression that women don't know what they're doing," said Christina Gabriel, a policy analyst for the Ontario Women's Directorate. "Women are also socialized to fit certain moulds so they are less likely to challenge an aggressive sales person."

Cars are one area where the price is elastic, and sales people are likely to drive a harder bargain if their customer is female. A 1990 study by the American Bar Foundation targeted 400 car dealerships in the Chicago area and discovered that, on average, women were charged a profit markup 40 per cent higher than men were charged. On a car that cost the dealer $11 000, the final offer to men was $11 362; to women, $11 504.

Such discrimination takes a hard toll, not just on women but on other businesses as well, Ms. Jones-Lee said.

"Women paying more for cars, for example, have less money to spend for computers, so if, in fact, there are industries which don't discriminate they ought to be up in arms about this practice because they're losing money."

Massachusetts now has a law prohibiting gender-based pricing. But Canadians have no such protection.

"None of it is legislated, none of it is regulated and none of it is controlled," said Joseph Lanzon, a consumer advocate who recently was a guest on an Ottawa radio show called *The Expert Hour* — a show that encourages listeners to call in. He said the majority of callers complained about gender-based price discrimination.

Mr. Lanzon advised listeners to avoid buying items packaged specifically to appeal to women.

In other words, buy blue razors instead of pink, and go to the men's department for that pair of gym socks.

But Anita Jones-Lee thinks ordinary citizens shouldn't have to tackle the problem on their own.

"I hope that ultimately some-one in the capital starts to take notice," she said. "Because it does have an insidious effect on the national economy."

On Aging

♦♦♦

BY

MAYA

ANGELOU

When you see me sitting quietly,
Like a sack left on the shelf,
Don't think I need your chattering.
I'm listening to myself.
Hold! Stop! Don't pity me!
Hold! Stop your sympathy!
Understanding if you got it,
Otherwise I'll do without it!

When my bones are stiff and aching
And my feet won't climb the stair,
I will only ask one favour:
Don't bring me no rocking chair.

When you see me walking, stumbling,
Don't study and get it wrong.
'Cause tired don't mean lazy
And every goodbye ain't gone.
I'm the same person I was back then,
A little less hair, a little less chin,
A lot less lungs and much less wind.
But ain't I lucky I can still breathe in.

Society Discriminates Against Teenagers

♦♦♦

BY

KATHLEEN CAWSEY

Every person is equal before and under the law and has the right to the equal protection and equal benefit of the law without discrimination and, in particular, without discrimination based on race, national or ethnic origin, colour, religion, sex, age, or mental or physical disability.
— The Canadian Charter of Rights and Freedoms

Today's teenagers are a minority that suffers major discrimination.

Most teenagers do not have the right to vote. They are not represented in government. They have a legal identity only through their parents. Teenagers are often paid less than adults even though the jobs are the same.

Many stores do not allow more than one teenager inside a time. Many restaurants do not allow teens to sit in certain areas.

Society continually concentrates on negatives when it comes to teenagers. Swarming, drugs, rowdy parties and teenage sex are all hot news items, while the contributions teens make to society with volunteer or community work are virtually ignored. Public workers such as bus drivers or police officers nearly always assume that a group of teenagers means trouble, and often provoke that trouble by acting in a domineering or authoritarian manner.

This discrimination goes against all that Canada and its Charter of Rights stand for. When you read about adult criminals you do not assume that all adults are evil. If you know an adult who does nothing but drink beer and watch TV you do not assume that all — or even a majority — of adults are couch potatoes.

Yet when a minority of teenagers are school dropouts or vandals, society seems to judge all teenagers by those few, typecasting them as apathetic and lazy or irresponsible hooligans.

Teenagers across the country take part in diverse community activities, in hospital or retirement-home volunteering, in environmental groups, anti-racist groups, religious groups, charity organizations, animal-rights groups, recycling groups. Meanwhile, they keep up their school work and take part in sports, musical events, school government and various clubs.

Yet much of society still has a negative view of teenagers.

What are adults afraid of? Why are responsible, respectable — if young — Canadians ostracized, ignored and shunned?

Young people have always been radicals, advocates of change. Sometimes they actually revolt against governments; but most of the time the rebellion remains in the home or the schools. Many adolescents feel the need to establish a separate identity from their parents, and they do this by rejecting their parents' values.

In most cases, this takes shape in small ways: long hair, loud music, earrings, and minor transgressions. However, some teens are more extreme and turn to drinking, vandalism or gangs to prove to society, and therefore to their parents, that they are no longer children.

Unfortunately, society sees the first form of rebellion as a symptom or a spark for the second, rather than as an alternative.

If a teen decides, for instance, to shave her hair, that teen's parents see images of crack and slimy boyfriends. These parents, instead of allowing their teen to win her small rebellion, clamp down harder and force her to find a different way to rebel — often a way that is more extreme.

People need to realize that a group of teens with long hair and leather jackets does not indicate a gang bent on mugging. While there will always be teens who go beyond simple rebellion, they are the exception and not the rule, just as adult criminals are the exception.

Society needs to recognize that it is not fair to apply stereotypes to teenagers. People tend to live up to the expectations of others, and once society starts recognizing the positive contributions of adolescents, teens will no longer need to find other means of gaining recognition and attention.

Adults need to encourage teenagers to channel their energies into worthwhile projects. By concentrating on the positives rather than the negatives, society would treat teenagers as valuable citizens rather than lost causes. History has proved that young people can and will change society. If today's teenagers see themselves as responsible, worthwhile citizens, Canada can only benefit.

Kingship

◆ ◆ ◆

BY

KIT

GARBETT

"**G**ather round," said the young man. "Call the villagers and those in neighbouring villages too. I've got something to say."

With some murmurs of "I've got dinner to cook" and other murmurs of "It sounds more interesting than cooking dinner," the villagers gathered round. The young man told them his idea.

"At the moment everybody does what they want, whenever they want as long as no-one else objects. If you, for example," he indicated a red-haired villager, "if you fancy fish you go fishing."

"I don't like fish," the villager said emphatically.

"Yes, but if you *did*, then you'd go fishing…"

"I've never liked fish. Too many bones…"

"OK, forget the fish. Take another example. If you fancied an apple, right…"

"Yes, I like apples. No bones. Pips and…"

"If you want apples," the young man continued, "You pick some in the forest, as you please. For the big decisions, though, that affect everyone, we all decide what to do, don't we?" He appealed to them. They all agreed.

"Now, I suggest that, instead of doing as we please, or having discussions, one person makes all the decisions. That will be me of course," said the young man hastily. "Everyone does what I say. It's a deal. I tell you what to do, in return you do what I say."

They thought this over for a while. It seemed fair, though the details would have to be worked out.

"What things do we have to do?" someone asked brightly.

"Everything," the young man replied. "Everything I say. If I fancy some fruit, whoever I tell has to get some apples."

"And if there's no apples, you get fish," the red-haired villager contributed happily. He could see what he was getting at.

"Almost," the young man agreed warily. His new subjects would need training. "Now, listen carefully. I'll be in charge and called the King. OK?" There was a general nodding of heads. The King hurried on as the red-haired man opened his mouth. "If any of you," he pointed a regal finger, "want an apple or anything, you ask my permission. If I say no, you can't. If I say yes, you can."

"Let's see if this is right," a tall villager began.

"Wait," the King stopped him. "Remember, you ask my permission before you do anything, so you have to ask my permission to speak, now go on."

"Right, so..."

"No. I meant go on and ask permission to speak," came the kingly rejoinder.

"I see," the tall villager beamed. "May I have permission to speak?" he petitioned.

"No," commanded the King. The tall villager sat down.

The King had further commandments. "Because I'm the King you do everything I say, and everything belongs to me. The fish in the river, the fruit on the trees, in fact the river, the trees, all the things on the land and everything I haven't thought of yet, all belong to me. You collect fruit, fish, wood, crops, animals. In return, I'll let you keep some for yourselves. You won't have to worry any more. That's fair enough, isn't it?" The villagers nodded cautiously. It seemed reasonable enough; was there a catch in it?

"Permission to speak, King?" asked the tall villager.

The King nodded graciously.

"It's OK so far," the tall villager continued. "But what if I don't do as you say, or do something without asking permission?"

The King smiled sagely. "That's easy. If you, or anyone else, disobeys my orders, then you, or whoever, will be killed. Quite messily. It's the simplest way. Any more questions?" The King smiled benevolently.

The tall villager was not satisfied. "I don't..."

"Permission," the King reminded him gently.

"Sorry, permission to speak please, King."

"Good, good." The King smiled at his subjects. "You're getting it now. Permission granted," he assented.

"Thank you, King," the tall villager continued. "This killing...I'm not sure about it. Will you do the killing yourself? After all, I am bigger than you."

"Good, I like this. You're anticipating. You show promise. You'll be knight or something. I don't get involved in the killing," the King explained. "I order someone else to do it. Special people do the killing. They're what I'll call an army."

"May the King have permission to tell us about armies?" asked another villager.

"Nearly right," said the King magnanimously. "I was going to tell you later. I shall need an army, of course. That's a lot of very strong men, called soldiers. The soldiers ensure that everyone obeys my orders. In return, the army protects you by fighting other armies. Of course, there aren't any other armies yet, but when they hear about us other people will copy us and have Kings and armies. Then we, or rather the soldiers, will fight them."

"May I have the King's gracious permission to ask another question?" said the tall villager.

"Very good indeed. You'll be a duke or something — go ahead," the King said, making a note to himself to keep an eye on the tall villager and have him beheaded at the first signs of competition.

"Pray, would the King tell us where he gets these noble soldiers?"

"That's the beauty of it," said the King. "Because you are all my subjects, I'll make the biggest of you soldiers. The rest work to support me and the army. In return, the army protects you from other armies and kings. Now that we've covered most things, the audience is closed. You have permission to leave. Except you," he added, pointing to a beautiful maiden.

"It has its points. I suppose," the red-haired villager muttered to the tall villager as they walked away. "But there's too many bones..."

"'My Lord Duke,'" replied the tall villager sharply. "Call me 'My Lord Duke,' peasant!"

The Convention
on the
Rights of the
Child

On June 1989, the three-master *Messager de la ville de Nantes* left the Breton coast to sail to Dakar, Fort-de-France and New York. On board were a dozen young people of different nationalities, aged from twelve to sixteen. They were joined by about fifty more on the island of Gorée, offshore from Dakar, and a third batch boarded the ship in the West Indies.

The children came from five continents, notably from the regions of the South. The voyage had a twin symbolic purpose: to retrace the old slave route from Africa to America, and to permit joint reflection on the main articles of a projected Convention on the Rights of the Child. At journey's end, the children went to United Nations headquarters in New York to present the Secretary-General, Mr. Javier Pérez de Cuéllar, with a petition they had written, requesting that the project, under consideration since 1959, should finally be ratified.

The Convention, which was eventually adopted by the General Assembly of the United Nations on 20 November 1989, fills a major gap, for hitherto children and young people had no legal rights before reaching the age of majority. This, moreover, is a time when growing numbers of them are deprived of parental protection and the security of a family, being literally thrown out onto the streets as a result of wars, famines, catastrophes or population movements. More insidiously, others suffer as a result of economic changes which tend to destroy community structures and marginalize the poorest and weakest sections of society.

It is important that the international community should give serious attention to the question of the global status of children and should do its best to find a morally and judicially satisfactory solution to the problems posed. In this respect, the voyage of the *Messager,* undertaken by youngsters many of whom had themselves endured years of hardship, has great symbolic significance, especially since the venture symbolically linked the misfortunes of today's endangered children with the suffering of slaves in a previous epoch. With the passing of the centuries, the struggle to affirm human dignity has passed into ever younger hands.

The Convention on the Rights of the Child has been described as a Magna Carta for children. It has fifty-four articles detailing the individual rights of any person under eighteen years of age to develop his or her full potential, free from hunger and want, neglect, exploitation or other abuses....

The Convention goes beyond previously existing instruments by seeking to balance the rights of the child with the rights and duties of parents or others who have responsibilities for child survival, development and protection, and by giving the child the right to participate in decisions affecting both the present and the future....

The inherent strength of the new Convention lies in its flexibility, its capacity to accommodate the many different approaches of nations in pursuit of a common goal. It has not evaded sensitive issues, but has found means to adjust to the different cultural, religious and other values which address universal child needs in their own ways.

Following is a summary of some of the main provisions of the Convention.

The Main Provisions of the Convention

Preamble

The preamble recalls the basic principles of the United Nations and specific provisions of certain relevant treaties and proclamations. It reaffirms the fact that children, because of their vulnerability, need special care and protection, and places special emphasis on the primary caring and protective responsibility of the family.

It also reaffirms the need for legal and other protection of the child

A student interviews James P. Grant,
UNICEF Executive Director, at the celebration
of the adoption of the Convention.

Students present their thanks to Audrey Hepburn, UNICEF Goodwill
Ambassador, James P. Grant, UNICEF Executive Director,
and Jan Martensen, Under-Secretary General.

before and after birth, the importance of respect for the cultural values of the child's community, and the vital role of international co-operation in securing children's rights.

Definition of a child
(art. 1)
A child is recognized as a person under 18, unless national laws recognize the age of majority earlier.

Non-discrimination
(art. 2)
   All rights apply to all children without exception. It is the State's obligation to protect children from any form of discrimination and to take positive action to promote their rights.

Best interests of the child
(art. 3)
All actions concerning the child shall take full account of his or her best interests. The State shall provide the child with adequate care when parents, or others charged with that responsibility, fail to do so.

Implementation of rights
(art. 4)
The State must do all it can to implement the rights contained in the Convention.

Parental guidance and the child's evolving capacities
(art. 5)
The State must respect the rights and responsibilities of parents and the extended family to provide guidance for the child which is appropriate to her or his evolving capacities.

Survival and development
(art. 6)
Every child has the inherent right to life, and the State has an obligation to ensure the child's survival and development.

Name and nationality
(art. 7)
The child has the right to a name at birth. The child has also the right

to acquire a nationality and, as far as possible, to know his or her parents and be cared for by them.

Preservation of identity
(art. 8)

The State has an obligation to protect, and if necessary re-establish, basic aspects of the child's identity. This includes name, nationality and family ties.

Separation from parents
(art. 9)

The child has a right to live with his or her parents unless this is deemed incompatible with the child's best interests. The child also has the right to maintain contact with both parents if separated from one or both.

Family reunification
(art. 10)

Children and their parents have the right to leave any country and enter their own for purposes of reunion or the maintenance of the child-parent relationship.

Illicit transfer and non-return
(art. 11)

The State has an obligation to prevent and remedy the kidnapping or retention of children abroad by a parent or third party.

The child's opinion
(art. 12)

The child has the right to express his or her opinion freely and to have that opinion taken into account in any matter or procedure affecting the child.

Freedom of expression
(art. 13)

The child has the right to express his or her views, obtain information, make ideas or information known, regardless of frontiers.

Freedom of thought, conscience and religion
(art. 14)

The State shall respect the child's right to freedom of thought, conscience and religion, subject to appropriate parental guidance.

Freedom of association
(art. 15)
Children have a right to meet with others, and to join or form associations.

Protection of privacy
(art. 16)
Children have the right to protection from interference with their privacy, family, home and correspondence, and from libel or slander.

Access to appropriate information
(art. 17)
The State shall ensure that information and material from a diversity of sources is accessible to children, and it shall encourage the mass media to disseminate information which is of social and cultural benefit to the child, and take steps to protect him or her from harmful materials.

Parental responsibilities
(art. 18)
Parents have joint primary responsibility for raising the child, and the State shall support them in this. The State shall provide appropriate assistance to parents in child-raising.

Protection from abuse and neglect
(art. 19)
The State shall protect the child from all forms of maltreatment by parents or others responsible for the care of the child and establish appropriate social programmes for the prevention of abuse and the treatment of victims.

Protection of a child without family
(art. 20)
The State is obliged to provide special protection for a child deprived of the family environment and to ensure that appropriate alternative family care or institutional placement is available in such cases. Efforts to meet this obligation shall pay due regard to the child's cultural background.

Adoption
(art. 21)
In countries where adoption is recognized and/or allowed, it shall only be carried out in the best interests of the child, and then only with the authorization of competent authorities, and with safeguards for the child.

Refugee children
(art. 22)
Special protection shall be granted to a refugee child or to a child seeking refugee status. It is the State's obligation to co-operate with competent organizations which provide such protection and assistance.

Disabled children
(art. 23)
A disabled child has the right to special care, education and training to help him or her enjoy a full and decent life in dignity and achieve the greatest degree of self-reliance and social integration possible.

Health and health services
(art. 24)
The child has a right to the highest standard of health and medical care attainable. States shall place special emphasis on the provision of primary and preventive health care, public health education and the reduction of infant mortality. They shall encourage international co-operation in this regard and strive to see that no child is deprived of access to effective health services.

Periodic review of placement
(art. 25)
A child who is placed by the State for reasons of care, protection or treatment is entitled to have that placement evaluated regularly.

Social security
(art. 26)
The child has the right to benefit from social security including social insurance.

Standard of living
(art. 27)
Every child has the right to a standard of living adequate for his or her

physical, mental, spiritual, moral and social development. Parents have the primary responsibility to ensure that this responsibility can be fulfilled, and is. State responsibility can include material assistance to parents and their children.

Education
(art. 28)
The child has a right to education, and the State's duty is to ensure that primary education is free and compulsory, to encourage different forms of secondary education accessible to every child and to make higher education available to all on the basis of capacity. School discipline shall be consistent with the child's rights and dignity. The State shall engage in international co-operation to implement this right.

Aims of education
(art. 29)
Education shall aim at developing the child's personality, talents and mental and physical abilities to the fullest extent. Education shall prepare the child for an active adult life in a free society and foster respect for the child's parents, his or her own cultural identity, language and values, and for the cultural background and values of others.

Children of minorities or indigenous populations
(art. 30)
Children of minority communities and indigenous populations have the right to enjoy their own culture and to practise their own religion and language.

Leisure, recreation and cultural activities
(art. 31)
The child has the right to leisure, play and participation in cultural and artistic activities.

Child labour
(art. 32)
The child has the right to be protected from work that threatens his or her health, education or development. The State shall set minimum ages for employment and regulate working conditions.

A
Small
Crime

◆ ◆ ◆

BY

JERRY

WEXLER

W hen he was nine years old, he was brought to the door by a policeman who kept one hand on his arm as if to stop him from running away. He had been caught writing with a crayon on a wall of the subway station, and his parents were expected to discipline him.

The rest of the day, he stayed in his room, waiting for his father to come home from the shirt factory. A slap on the face, he thought, perhaps that's all I'll get. And maybe no allowance for the coming week. Still, he could not help but be apprehensive.

At five-thirty, he heard the front door open. His mother was talking to his father. They talked for a long time, much longer than he felt was necessary. Then the family ate supper. He was not invited and he felt that his punishment had already started. This saddened him greatly because he enjoyed eating supper with his father and telling him about the day's adventures. He tried to pass the time by reading through his comic books, but he was anxious and could not follow one through from beginning to end.

At seven o'clock, he heard the television come on as the family sat down in the living room. Every so often someone changed a channel. By eight-thirty night had fallen, and he felt more alone than he had ever been at any time in his life. He looked out into the garden behind his room. He could see the outline of the young tree his grandfather had planted a month before. It was beginning to sprout leaves, and he tried to decide how many it would have that summer. He decided on fifteen. That knowledge made him feel better. By nine o'clock he was feeling drowsy and had lain down on his bed.

Shortly after, his father came into his room and sat down on the bed. The boy sat up, putting his feet over the side of the bed. His father looked at him, then turned away clasping his hands together. They were sitting side by side. The boy looked down at the floor the whole time.

"I remember when I left Romania," his father said, "I went to the train station in my town and had to sit in a special section for people who were emigrating from the country. There were many other people sitting there, and I remember how funny it was that we all looked the same with our best clothes and old suitcases almost bursting with clothing. But what I remember most of all were the walls. It seemed that every person who had ever sat in that part of the railway station had written on the wall his name, his home town and the new place to which he was going. I spent a long time reading all the names on the wall. Many towns were represented, and I even recognized the names of many people with whom I had once been friends. And so you know what I did? I took out my pen, found a clear space on the wall and wrote my own name, my home town and the date.

But you see, I did not write on the wall out of mischief. It was my own way of saying, *this is who I am — now I am ending an old life and starting a new one.* Perhaps one day you will have the same reason. But as long as times are good, and we are welcome here, there are other and better ways of letting the world know who you are."

He put his hand on the boy's shoulder as if to say, "Don't worry, everything is all right." Then he rose slowly and said: "Be good to your mother and me and leave the walls alone." Then he left the room.

The boy lay back on his bed. Ordinarily it took him only two or three minutes to fall asleep. That night he lay awake for almost half an hour.

Plenty

◆◆◆

BY

JEAN

LITTLE

I have plenty of everything
       but want.
I try to imagine hunger,
Try to imagine that I have not eaten today,
That I must stand in line for a bowl of soup,
That my cheekbones angle out of my hollowed face;
But I smell the roast in the oven.
I hear the laden refrigerator hum.

I think of people whose walls are made of wind.
I stand outside in the cold.
I tell myself I am homeless and dressed in rags;
But my shiver lacks conviction.
I stand in fleece-lined boots and winter coat.
Home is a block away.

I leave my wallet at home.
Pretending I have no money,
I walk past stores and wish.
"I have no money, no money at all, no money — "
I turn my head in shame as I pass the bank.

I pay for a parcel of food. I gather clothes.
I adopt a child under a foster parent plan.
I do what I can. I am generous. I am kind —

I still have plenty of everything
but want.

Walking Tall, Talking Tough

◆◆◆

BY

FIONA SUNQUIST

On a chilly morning in January, an elderly jeep wheezes to a halt in front of the wildlife warden's office in India's Nagarhole National Park. Four khaki-clad guards leap out, salute warden K. M. Chinnappa, then drag three disheveled, handcuffed men from the back of the vehicle. The guards step back quickly and come to attention, leaving Chinnappa room to pace around the prisoners.

The warden is a warrior, a fighter. One can see it in his eyes, a haughty, hooded look that probably describes his thoughts better than any words. His lip curls with disdain as he stares at one of the handcuffed men, examining him as if he were something unpleasant on the bottom of a shoe. One captive falls to his knees and crouches on the ground pleading for mercy. Chinnappa responds with machine-gun-like bursts of questions without waiting for answers. His anger is so intense it would be easy to assume that the prisoners had committed a personal assault on his family.

The handcuffed men are suspected of being the front team for a group of ivory poachers. They maintain they are on a sightseeing trip, but Chinnappa coldly weighs the evidence. They have a large stockpile of food and, say the guards, a suspiciously well-organized camp. Within an hour, Chinnappa ships the group off to jail.

These are the front lines of the fight to save India's wildlife. While the Western world spends millions of dollars on captive breeding programs, poorly paid wardens like Chinnappa risk their lives daily in the world. If there are heroes in the conservation movement, they must include these soldiers in the field. And perhaps the best example of what it takes to get the job done in the Third World is the enigmatic warden of Nagarhole, known among his fans as "India's Field Marshal of Conservation."

Aggressive, rude, even

arrogant at times, the warden is a man of inestimable integrity who does what he wants largely on the strength of his sometime overbearing personality. "Plain and simple, Chinnappa has what it takes to protect wildlife in the real world," says Praveen Bhargav, a longtime Nagarhole visitor. "He leads by example and has high standards for himself and others."

At six-foot-four, Chinnappa stands head and shoulders above the diminutive tribal people who live in Nagarhole. His ailing back exaggerates his stiff posture, and his downright skinny frame reflects an abstemious life-style; he does not drink, smoke or eat meat. His home is spartan. As if to compensate for his ascetic side, he has an undergraduate-style sense of humour and revels in bad puns and ribald jokes.

For more than 20 years, Chinnappa has struggled to protect this park on an annual salary of less than $2000, winning widespread acclaim and a gold medal from the chief minister of the state of Karnataka. But his uncompromising style has won him as many enemies as friends, and when his political foes succeeded in removing him from his post for a year, his importance became graphically clear as Nagarhole quickly deteriorated. "It was an illustration of how vulnerable a park is without good protection," says Ullas Karanth, a biologist with the Center for Wildlife Studies in Mysore.

For Chinnappa, the reward is not so much in acclaim as in his work. "The pleasure comes from the animals, especially elephants," he says. "Just watching them, I've learned a lot about their lifestyles. With understanding came respect, for their intelligence, family life and sense of humor. They are beautiful." In a sudden dark mood shift, he adds, "It is a terrible thing to kill such an animal."

Chinnappa spends his days stemming the tide of poached animals, forest produce and sandalwood from Nagarhole. His job, he believes, is to guard against anything and anyone who threatens the sanctity and integrity of the park. And the threats are many: Nagarhole contains valuable timber and wildlife that can be eaten or sold. Three-quarters of the park's boundaries abut agricultural land or coffee plantations, which increases potential entry points for poachers. Forest fires, unruly tourists and an eclectic assortment of other crises add to his headaches.

Nagarhole sticks out like a thumb from a jigsaw of four interconnected parks and sanctuaries in the Mysore region of southern India. Rich in wildlife, the area is one of the last strong-holds of the Asian elephant. At least 2000 of these endangered giants still roam through the four neighboring protected areas. Nagarhole is also one of the few places where it is possible to catch a glimpse of another rare heavyweight, the massive wild oxen known as gaur. With their heavy horns and rippling mus-cles, gaur stand 6 feet [183 cm] tall at the shoulder and weigh as much as a ton.

Other smaller species such as spotted deer, sambar deer, wild pigs and langur monkeys abound. Predators include tigers, leopards and wild dogs. The region's more than two hundred bird species include herons, cor-morants, eagles, gray jungle fowl and such uncommon creatures as the rocket-tailed drongo and the blue-bearded bee eater. The biomass — or weight of animals per square kilometer — rivals that of many East African parks. But, unlike the animals of the open savannas of Africa, much of Nagarhole's wildlife is hard to see, hidden in dense forest that houses some of the most valuable trees in the world. Ebony and rosewood used for carvings and intricate inlay work, teak for furniture and veneer, and sandalwood prized for its scent mingle with other high-priced timber trees.

In countries like Indonesia, Brazil and Zaire, whose forests hold much of the world's biodi-versity, government budgets are stretched to bursting, and, not surprisingly, park protection is low on the list of priorities. Wildlife departments are often poorly staffed and ill-equipped to police the areas in their charge, and the pressure on wardens is so intense that some give up....

At Nagarhole, Chinnappa simply plugs away, often working seven days a week in a job that is demanding, difficult and danger-ous. "We know how lucky we are to have him," says Karanth. "It is people like Chinnappa who really protect biodiversity. Without them, all the theoretical calcula-tions of inbreeding and minimum park size are just so much paper."

In the forest, few details escape Chinnappa's attention, and he reads animals tracks as easily as some people read a newspaper. Combining the mind of a good detective with the skills of an experienced naturalist, he is quick to follow up odd or sus-picious reports. On a typical day, a report of a 400-pound [180 kg] sambar deer killed by a tiger near the western park boundary piques his interest immediately. Bellowing orders to the trackers and guards, he folds his long legs into the jeep and races off to inspect the site half an hour away.

At the scene, everyone stands

back as Chinnappa walks lightly around the dead animals, carefully inspecting the ground. He crouches for a closer look at the leaf litter, carefully moving the leaves aside, looking for clues — hair, footprints, drops of blood. Oblivious to the swarming flies, he leans over the bloating carcass, pointing out the lack of bite marks or blood on the body, and the lack of a sign of a struggle. To Chinnappa, it is already clear how the deer died — and it had nothing to do with a tiger. He motions to two of the guards to turn the carcass over, then shakes his head in the Indian gesture of affirmation. There is a small hole just behind the right shoulder. The animal has been shot. And this time the poachers have gotten away.

Chinnappa has recruited some of the best forest guards available and organized them into highly effective anti-poaching patrols. "The guards need to be tough and willing to face risks because poachers often fire back at you," says Krishna Prasad, a founding member of WILPEG, a Bangalore wildlife society.

The guards camp out in the jungle for a week at a time, hiking to reach inaccessible areas. Their camps consist of a simple grass roof over a living area and a tree platform to escape being trampled by the elephants they are trying to protect. "It can be a dangerous job if you do your work sincerely," says Chinnappa. "Sandalwood smugglers and elephant poachers are usually armed, they operate in large groups of 20 to 25 people and they don't want to be caught." They have killed at least three guards in Nagarhole in two decades.

Just as daunting as coping with lawbreakers, for Chinnappa, is dealing with local politics and customs. Wardens and guards of Asian national parks are usually poorly paid, and it is not uncommon for guards to succumb to the temptation to supplement their meager salaries with bribes from poachers, or even to take up poaching themselves. Not Chinnappa, who has been known to make coffee growers drive their loaded trucks an extra 70 miles [112 km] around the park rather than risk hit-and-run accidents with wildlife.

Fans like Bhargav believe that much of Chinnappa's success as a warden is due to his policy of aggressive protection with no exceptions. "Chinnappa believes that the same rules should apply to everyone," says newspaperman R. Subramanium, correspondent for the *Times of India*. This approach is not always well received in India, where issues of caste and class have traditionally determined people's privileges and social standing.

Chinnappa's egalitarian attitude toward law enforcement has led to many skirmishes with officials who are accustomed to pulling rank and getting results. For instance, the park gates close at six o'clock and no one is allowed to drive through after closing time. Local lore has it that several senior officials of the police and forest department are among the many people who have been turned away for arriving too late in the day without making prior arrangements.

In 1987, the police arrived at Chinnappa's front door with an arrest warrant. A local coffee planter had been shot and killed by a forest guard, and, as the guard's boss, Chinnappa was accused of orchestrating the killing. Briefly jailed and publicly humiliated, Chinnappa wrote long letters to his superiors in the Indian Forest Service. "He very nearly gave up," said Praveen Bhargav, "and there was a lot of debate as to whether he would return to his job when it was all over."

Chinnappa's supporters rallied. They located a lawyer and wrote letters to everyone from the Prime Minister on down. "I never knew I had so many friends," says Chinnappa wonderingly at the outpouring of support. Later acquitted of the charge that had been trumped up by his enemies, Chinnappa was released and transferred to another job where he stayed for a year. In his absence the park languished. "It was frightening how fast Nagarhole began to go downhill while Chinnappa was in prison," remembers Charles Fernandes, a research assistant with the University of Agricultural Sciences in Bangalore.

"The guards would go out for night patrol at eight o'clock and come back an hour later. No one cared what they did. All night long you could hear gunshots in the park. No one followed up on them, and tourists who inquired were told that they were fireworks." Wildlife along all the main roads began to disappear as illegal hunting continued apace. Lobbying by Chinnappa's friends and long-term park aficionados finally succeeded in getting him reinstated. With Chinnappa cracking the whip again, protection has improved.

When Chinnappa first came to work in Nagarhole 21 years ago, thousands of people were living in the forest and poaching was rampant. The *hadlus*, or swamps, were rice fields ringed with villages. As people were resettled, the hadlus reverted to their natural state and the wildlife gradually returned. Seeing this healing process take place has been one of Chinnappa's greatest rewards. "Today, when I watch a group of elephants or a herd of gaur in the swamp," he says, "I remember how it used to be, and it gives me a feeling that all things are possible."

The Thief Who Kept His Hands Clean

♦ ♦ ♦

BY

CAROL KENDALL &

YAO-WEN LI

L ong years ago in south China there lived a magistrate called Chen whose wisdom equalled his love of justice. He was admired and revered by all the honest people under his rule.

One night there was a great robbery in the district. The constable, eager to win the admiration of Magistrate Chen, quickly swept up every possible suspect in the neighbourhood and crammed the lot of them into the courtroom. He then set himself to questioning each in turn, asking the whereabouts and the wherefores and the whens and whats, until he had amassed so many twists of fact that he was quite entangled in them. At last he had to admit defeat and, as on many another occasion, he turned the hopeless jumble over to the magistrate.

At the preliminary hearing, Magistrate Chen heard the charges, but instead of questioning the suspects, he announced, 'Upon the eastern hill in the Temple of the Great Buddha there is an old bronze bell which will tell us who the robber is.'

Thereupon he sent the constable's men to transport the bell from the temple to the court, and gave orders for a blue cotton canopy to be spread on poles above it. The night before the trial he himself set firepots in the four quarters round the bell and lit them. When the fires finally died away, he slowly lowered the blue canopy until the bell was completely hidden in its folds.

Because Magistrate Chen's judgments were famous, people thronged to the courtroom the next day to attend the trial. There was scarcely room for them and all the suspects too.

Without preamble the magistrate addressed the suspects. 'This old bronze bell from the Temple of the Great Buddha has powers of divination beyond those of man or magistrate,' he said. 'Ten thousand innocent people may rub it and no sound will be heard, but let one thief touch his hand to its side, and a clear peal will sound out his guilt for all the world to hear.' He paused and looked intently at the suspects. 'In a few moments I shall ask each of you to put his hand under the blue cover and rub the bell.'

Solemnly, the magistrate bowed his head and made a prayer in front of the bell. Then, one by one, he led the suspects forward and watched while they put their hands under the cover to rub the bell. As each suspect turned away without the bell's having sounded, the crowd gave a little sigh, but when the last man had passed the test, there was a restless stirring. The bell had failed!

But the magistrate clapped his hand on the shoulder of the last suspect.

'Here is the thief we are looking for,' he declared.

A stir of outrage spread over the crowd.

The accused spluttered. 'Your Honour! When I rubbed the bell, there was no more sound than when all the others did the same thing! How can you so unjustly accuse me!'

'He's right!' somebody in the crowd shouted.

Indignant cries rang out on all sides.

'Unfair!' 'Injustice!'

The magistrate, unperturbed, gave his beard a tidy stroke. 'Remove the cover,' he said to the constable.

The constable did his bidding.

There was a gasp of amazement around the courtroom, and then the crowd fell silent as they began to understand what had passed. The gleaming bell they had seen brought to the court was now black with soot — save round the rim where many innocent hands had rubbed through the grime to the gleaming bronze.

'Truly this old bell has powers of divination,' Magistrate Chen said. 'Truly, it uttered no sound when innocent people rubbed it. And it uttered no sound when the thief put his hand under the cover, for only the thief was afraid to rub the bell for fear of its revealing peal. Therefore, only the thief brought his hand out from under the cover free of soot.'

All eyes came to focus on the shamefully clean hands of the thief, and a murmur of admiration swelled from the crowd.

Truly it was a magic bell. And Magistrate Chen was something of a magician.

The Orangutan
Who Believed In
Simple Solutions

♦♦♦

BY

ALAN

NEIDLE

An ancient orangutan who enjoyed great renown as a distinguished jurist often pondered the nature of animal life in the jungle. What he saw as he peered down from his nest in one of the tallest fruit trees deeply disturbed him. Snakes eating birds. Tigers chewing up pigs. Beetles poisoning grasshoppers. Eagles snatching monkeys. Crocodiles crunching turtles. Shrews gobbling up mice.

"There must be a way to put an end to all this chaos and injustice," the orangutan jurist said to himself. "We need a law that will bring happiness to everyone."

And so the orangutan convened a committee of apes, all eminent jurists, to assist him. "What's needed," he told his colleagues, "is a simple concept. It should be a simple straightforward rule. If we create a lot of complex provisions, everyone will argue over them and no one will be happy — except the lawyers."

After a lengthy deliberation the committee came up with the following doctrine: "Animals may only eat smaller animals." The idea was that everyone would be most likely to comply with a regulation which permitted doing what was easiest.

The orangutan jurist was pleased with his work. "We have conceived," he declared, "a simple solution. Everyone will be able to understand it." All the other jurists joined in offering their praise. "It's not only simple," said one "it's positively eloquent."

After the new doctrine was promulgated, the crocodiles and the anteaters were, not surprisingly, among its strongest proponents. But some of the small cats complained that they didn't want to give up eating jungle pigs. The mongooses, who killed large snakes, also protested. But the warrior ants were the most vehement in their denunciation. They devoured everything, big and small, that didn't get out of their way. As predicted, everyone understood the new rule, but only the largest animals respected it. Chaos in the jungle continued.

The old orangutan reconvened his committee of eminent jurists. "My friends, do not be discouraged. Great new ideas are not always readily accepted. Perhaps through force of our collective wisdom we can devise an even better rule."

The committee explored many proposals, but none was satisfactory. Finally, when they were all quite dejected, the orangutan jurist clutched his brow and, after a profound silence, spoke these words, "Animals may only eat larger animals." The jurist explained that there would be much less killing if the regulation permitted animals to do only what was difficult.

The other jurists were deeply impressed. "This is one of those inspired ideas," said one of them, "that nobody's ever thought of before. But once you hear it, you wonder how everyone else could have missed it."

The warrior ants received the new law with enthusiasm. So also did the vampire moths who sucked blood from tapirs and large rodents. But most of the other animals were outraged. The leopards threatened to climb into the jungle canopy and teach the orangutans a thing or two about realistic rule-making.

For the third time the old orangutan convened his committee of learned jurists. "There must be a way," he told them, "to find a simple solution — if only we think hard enough." The committee conducted a searching review of all animal proclivities. It analyzed, synthesized, and conceptualized. Brows were deeply furrowed. At last, after prolonged ratiocination, the orangutan jurist came forth with the following formulation: "Animals may eat other animals."

The other jurists were beside themselves with praise for their leader. "It just goes to show," said one, "that if you try hard enough you can think of anything."

The latest law was applauded by everyone. And so the distinguished orangutan jurist basked in universal praise. Although chaos continued throughout the jungle, the jurist took no notice of it. He saw only one thing. All the animals were following his simple law.

What's the Verdict?

You're the judge in these courtroom quizzes

♦ ♦ ♦

BY

TED LeVALLIANT and

MARCEL THEROUX

1 Flash Fire

John negligently caused a flash fire in a restaurant. An employee activated an extinguisher, which caused a hissing sound. On hearing the sound, a customer shouted that gas was escaping and that there would be an explosion. The customers stampeded and Harry was injured in the stampede. Harry sued John for damages. *Does Harry succeed?*

2 The Unsuccessful Pickpocket

Connie put her hand in the victim's pocket. The victim grabbed Connie's wrist while her hand was in the pocket. The charge is attempted theft. There was no money in the pocket. *Is Connie guilty?*

3 Hair Today and Tomorrow

A hair removal clinic offered to permanently remove facial hair. It advertises that the results are guaranteed. In response to the ads, Shirley went to the clinic and paid for the treatment. Shirley's hair loss was not permanent and she sued for breach of contract.
Did Shirley win?

4 You Can Rely On Us

The accountants prepared financial statements for a client so that he could obtain financing from the bank. The accountants were negligent in the preparation of the statements. The bank had no relationship with the accountants, but the bank relied on the statements. The client went bankrupt. The bank sues the accountants.
Does the bank succeed?

5 The Joke's On You!

Morris had been drinking when he entered the bank. "I have a .38 in my pocket," he said to the teller, "and I want all your money." The teller set off a silent alarm. But when she handed Morris the cash, he said he had been joking all along. He left the bank empty-handed and was arrested.
Is Morris guilty of attempted robbery?

. . . .

1 Flash Fire

Trial Court Decision
Harry succeeds.

Appeal Court Decision
Harry does not succeed, because the damages were not reasonably foreseeable and were accordingly too remote to justify recovery.

> *Bradford versus Kanellos,* decided by a panel of three Judges of the Ontario Court of Appeal, in December, 1970, decision rendered by Judge Schroeder.
> *Formal legal citation:* [1971] 2 O.R. 393 (*Canada*)

2 The Unsuccessful Pickpocket

Trial Court Decision
Connie is guilty.

Appeal Court Decision
Connie is guilty. The fact that the theft was impossible is not a defense to a charge of attempt.

> *R. versus Scott,* decided by a panel of three Judges of the Alberta Supreme Court, Appellate Division, in November, 1963, decision rendered by Judge MacDonald.
> *Formal legal citation:* [1964] 2 C.C.C. 257 (*Canada*)

3 Hair Today and Tomorrow

Trial Court Decision
Shirley does not win.

Appeal Court Decision
Shirley won, because the advertisement was an offer. Shirley accepted the offer. There was a contract.

> *Goldthorpe versus Logan,* decided by a panel of three Judges of the Ontario Court of Appeal, in March, 1943, decision rendered by Judge Laidlaw.
> *Formal legal citation:* [1943] O.W.N. 215 (*Canada*)

4 You Can Rely on Us

Trial Court Decision
The bank does not succeed.

Appeal Court Decision
The bank succeeds, because accountants are liable to any person who might reasonably be expected to rely on their statements.

> *Haig versus Bamford,* decided by a panel of nine Judges of the Supreme Court of Canada, in April, 1976.
> *Formal legal citation: [1977] 1 S.C.R. 466 (Canada)*

5 The Joke's On You!

Trial Court Decision

Morris is guilty of attempted robbery.

Appeal Court Decision

Morris is not guilty of attempted robbery, because the evidence was equally consistent with no criminal intent as with intent that was abandoned.

> *Regina versus Mathe,* decided by the British Columbia Court of appeal, in April, 1973, decision rendered by Judge Maclean. *Formal legal citation: [1973] 4 W.W.R. 483 (Canada)*

Justice is often symbolized
in art by a blindfolded woman holding
a sword and scales.

Court Jesters:

Funny Stories from Canada's Courts

◆◆◆

BY

**PETER V.
MacDONALD**

Anyone who's run afoul of the law would be well advised to heed the words of the old saying, "He who acts as his own lawyer has a fool for a client."

Robert J. Lane, a lawyer, in Shellbrook, Saskatchewan, provides proof positive that truer words were never spoken. Bob sent me a court-certified transcript of the arraignment — in Provincial Court in Prince Albert — of a fellow who represented himself on a charge of breaking and entering.

After the charge had been read to the accused, the court clerk informed him that he could elect to be tried in that court or in the Court of Queen's Bench, and then the judge explained this to him. We pick it up from there:

The Court: Do you understand it now, or would you like further explanation?
Accused: I understand it.
The Court: Do you feel that you're prepared to elect and plead today, or do you want me to adjourn it so that you can take advice?
Accused: I'll take it.
The Court: You want to elect today? Who do you elect?
Accused: You.

The Court: And how do you plead, guilty or not guilty?

Accused: I plead not guilty.

The Court (to the prosecutor): How many Crown witnesses will there be?

Accused: Only one — the one I broke into the store with.

. . . .

In another western case, a chap who acted for himself in fighting a robbery charge asked the man who'd been robbed: "Could you see my face clearly when I handed you the note?"

. . . .

Corpus delicti is a Latin term that's tossed around a lot. It does not mean a corpse, as many people believe, but "the body of a crime," a substantial fact that proves a crime has been committed.

Bob Prince, a lawyer in Yarmouth, Nova Scotia, heard two men plead guilty to a charge of a theft of a pig. After hearing the facts, the judge asked what had become of the pig. "We ate it," one of the men replied.

The judge smiled, then said to the prosecutor, "It's fortunate for you that the accused entered guilty pleas. Otherwise, the Crown would have had no *porcus delicti.*"

. . . .

This lawyer should take a few days off work:

Q: And what did he do then?

A: He came home, and the next morning he was dead.

Q: So when he woke up next morning he was dead?

. . . .

This lawyer probably needs a rest, too.

Q: Where was the damage to your car?

A: He crumpled my right front fender.

Q: Which right front fender?

When you ask a stupid question, you're spoiling for a snappy comeback. For example:

Q: Could you see him from where you were standing?

A: I could see his head.

Q: And where was his head?

A: Just above his shoulders.

. . . .

And try this one on for size:

Q: So you say that the stairs went down to the basement?

A: Yes.

Q: And these stairs, did they also go up?

. . . .

This witness doesn't waste words:

"What do you do for a living?"

"I help my brother."

"What does your brother do?"

"Nothing."

. . . .

Lawyers learn early that you win some, you lose some. Ottawa lawyer William T. Green, Q.C., remembers the time he walked into court just seconds after a colleague, Dave Casey, had gone down to defeat. Green hadn't heard the verdict and, as Casey was preparing to leave, he asked, "How'd you do, Dave?"

"Not bad," Casey replied. "We came second."

Partners in Crime Prevention

♦♦♦

BY

LUIS

MILLAN

Sordid stories of domestic violence, senseless assaults, gang shootings, and brutal drug killings occurring in cities across Canada have become depressingly common in recent years. So much so that millions of Canadians, particularly women and the elderly, refuse to walk alone at night for fear of becoming a victim.

Clearly, the perception among Canadians living in urban centres is that crime, especially violent crime, is spiralling rapidly out of control.

Statistics reveal that the public is correct in assuming that there is a surge in criminal activities. After small increases of 1% or 2% a year for the past several years, the crime rate jumped by 6.2% in 1990 over the preceding year. But, whether this is merely a blip or the start of a long-term trend is still too early tell.

What is certain is that crimes involving violence have skyrocketed alarmingly over the past three decades. Between 1975 and 1989, the violent crime rate increased by 59%, from 597 to 948 incidents per 100 000 population. In spite of this trend, Canadians greatly overestimate the amount of violent crime for it comprises less than 10% of all reported crimes.

Police and criminologists feel they already know what causes crime — unemployment, poverty, an unresponsive educational system, and the breakdown of both families and communities. What they're trying to figure out is how to win the front-line battles against crime.

Many experts say that the only way to go is to hire more police officers and give convicted criminals tougher sentences. More policing is needed, they argue, because there are more criminals to catch. And, while prisons may not rehabilitate criminals, they do keep them off the streets for a while — making it the best deterrent to crime.

This position has the backing of the public. Poll after poll indicates that large numbers of Canadians have lost faith in the criminal justice system, believing that the courts are far too lenient.

But, is this expensive, punishment-oriented approach really the answer? Canada already has the third-highest level of incarceration in the world, jailing 109 people for every 100 000 in the population. Only the United States and Northern Ireland put more people in jail on a per capita basis.

While the public and politicians up for election call for heavier sentencing, several police forces are switching to something called community policing.

Community policing turns its back on just responding to emergency calls or having officers cruising neighbourhoods in their cars. Officers working in this system spend as much time with people in their neighbourhoods as chasing criminals. The key to making this system work is getting the community and the police to work hand in hand, say police officers.

Community policing has, in fact, become the new buzzword among those working in the criminal justice system. While many believe it's not tough enough on criminals, community-based policing has proven to be an effective tool in battling crime.

The Halton (Ontario) Regional Police has been using community policing since the early 1980s. Montreal's police force began implementing the concept a couple of years ago, and Toronto's officers in blue might be practising it in years to come.

In Halton, a mostly suburban municipality of 280 000 people, 20 kilometres west of Toronto, the 500-member force has become a model for other forces around the world. Divided into neighbourhood patrols, police officers are encouraged to participate in local activities. Moreover, contrary to the way that most police forces operate in North America, Halton's police spend half of their time on preventive policing; teaching people how to protect their homes and themselves from becoming crime victims. Just as important is the fact that local residents have a say in which the policing priorities should be.

In Montreal, a similar program was launched five years ago, but on a smaller scale, involving only 300 officers. But, given the success of the program, it will expand, says Montreal crime prevention police constable Denis Lemieux. In the future, Montreal citizens can expect to see more police officers walking the beat or patrolling their neighbourhood on bicycles, as is now done in Toronto.

"The fight against crime can't just be fought by the police," explains Lemieux. "Establishing a closer relationship with the public is a must so that we can work together. We want to eliminate the concept that cops can only be reached in case of emergency."

Anti-drug campaigns, assisting neighbourhoods to organize "Neighbourhood Watch" programs, giving the public at large tips to minimize the threat of being a victim — all these are things more Canadian police forces are doing in their fight against crime.

Other cities in Canada are taking a different tack. In Vancouver and Winnipeg, for example, municipal politicians along with different interest groups are examining ways that urban design can be improved. By doing little things such as placing bus shelters in safe areas, improving lighting in public parks and parking garages, asking people to keep their porch lights on from dusk to dawn, the opportunities for crimes can be reduced, the reasoning goes.

Implementing community crime prevention programs, however, hasn't always been a cakewalk, admits Lemieux. Like other modern, multicultural Canadian communities Montreal is plagued with tensions between the police force and different cultural groups.

Convincing police officers and citizens to work together in good faith can be difficult at times for it means forging new attitudes about each other. And, old habits and attitudes die hard.

My Theme

◆◆◆

BY

HANS

JEWINSKI

the chase is real
even if my theme
music doesn/t cut
in as the engine
makes the car jump

the chase is real
even if the dispatcher
repeats "use caution 5107"
"keep up your location 5107"
in most un-hollywood fashion

the needle climbs
and the tires squeal
through long blind curves
and quick stops

i can only see the
tail-lights "give us
a licence number 5107"
"keep up your location 5107"

westbound on queen st
in rush-hour traffic
the mustang is running
all the lights — then
the wrong way on a one way
through a busy parking lot
over a curb grass through
a hedge and a pedestrian
nearly stays in the way

"let him go 5107" yeah
follow policy 5107 "disregard
5107" "your location 5107"
don/t get any bystanders hurt

"your location 5107"
i turn off my red light
i lower my highbeams
and tell the dispatcher

i/m discontinuing the chase
"last direction of travel 5107"
last direction of travel 5107

don/t let it get personal
forget it — just another chase
"your location…"

my guts tied in a knot
i pull to the side of the road
and turn off the blaring radio
just another chase and the theme
music in my veins boils over
and i vomit all over the dash

Young Offenders Act, Pros & Cons

◆◆◆

BY

LORI

MARTIN

The Professionals are split on the Act's effectiveness

As a criminal lawyer, Percy Smith believes that present day offenders are worse off than those convicted under the older, stiffer act, because society no longer makes young offenders accountable for their actions.

Releasing the youth's name and publishing his or her crimes would serve as a deterrent, the Vancouver area lawyer explained. "If someone is going to behave in an antisocial fashion, I don't see why in one age group it should be kept secret and in others, it isn't," he noted.

In Smith's general law practice in Tsawwassen, near Vancouver, he finds the YOA prevents offenders from being responsible and accountable for their actions....

Describing the YOA as a "revolving door," one Metro Toronto police officer suggests that the YOA place offenders under supervision for longer periods of time. The trend has been to charge and convict and administer much shorter sentences, resulting in many more charges and convictions.

Bob Dzus, officer of the youth bureau at 52 division, says today's YOA is a revolving door in terms of supervision. "Kids today aren't afraid of the sentences they receive. They know they're not going to go to a jail-type facility. They're going where you get three square meals, watch TV, do a few things and then sit down and listen to a counsellor."

His solution? Force offenders to receive a basic education by putting them in a controlled environment where they have to do it. "A lot of these kids need someone to stand over them and make sure things get done," he says.

According to a Lethbridge, AB city police officer, the YOA succeeds with first time offenders, but misses the mark on repeaters.

"The people it's designed to work with — first time offenders — it deals with them effectively. But when you get into the kids who are habitual offenders, the system doesn't address the seriousness of the crime," says

Detective Ian Cameron.

He believes the YOA is better than the Juvenile Delinquent Act but recommends dropping the age maximum from 18 to 16. "If the 16, 17, and 18-year-olds appear in adult court, the YOA would be dealing with the young people it was intended to deal with," he says.

He'd also cancel the Act's Alternative Measures Program, which has a first-time offender completing some kind of community service. "If a young person needs a break, that discretion should be given to the police officer on the street," says Cameron, a 14-year veteran of the police force.

A Peterborough, ON criminal lawyer agrees.

"Back in the old days, before the YOA, the police would exercise a little more discretion right at the scene," says lawyer Dick Boriss. "If a young person got himself into trouble, they'd give him a blast, take him to the station and give him a lecture. Then they'd talk to his parents and let him go."

Boriss says now more offenders end up in court than they did before the act was in power.

A large part of his practice is young offenders, and Boriss believes young offenders "are singled out. And so I think there's more of a tendency for the police to say 'look, if there's an offense here, we're going to charge them under the YOA.'"

Boriss attributes the YOA's success to the interpretation of the presiding judge. "In our particular area, we have one pretty tough judge who handles young offenders and his theory is basically 'hit 'em hard the first time and they won't come back,' which is basically contrary to the philosophy of the young offender legislation. He treats young offenders in much the same manner as adults."

Boriss speaks with 20 years experience in recommending the age limit be lowered to 14 and up for adult court, keeping the YOA for cases involving 12-13 year olds.

"Under 14 would be a more appropriate age these days," he said.

Judge Doug Morton administered youth justice under both acts during his years on the bench. By the time he retired, Morton had defended young offenders as a lawyer under the Juvenile Delinquent Act and had enforced both the JDA and the YOA on the bench.

What does he think? Starting the YOA at the minimum age of 12 was a main concern when the act was introduced, he remembers. Because children under 12 could not be charged, some members of the system were afraid that youngsters would be recruited by older offenders to take part in illegal activities. But, overall, Morton thought the act was adequate.

The YOA incorporates much of the regular criminal code, Morton noted, which was considered a positive step. The retired judge reflects on the fact that, in his opinion, many of the kids had lousy parents. "One thing I always said was that I would treat these kids right."

For the John Howard Society in Halifax, NS, the young offender program is a roaring success. Robert MacDonald, executive director of the Halifax branch, said his organization has been running a five-part program since 1986. One area, alternative measures, involves using trained volunteers to act as mediators between the offender and the victim. "It's working. It's keeping young people out of the court system. A lot of them are not coming back," said MacDonald.

Give them a chance!

If Claire Culhane had her way, all kids would be given a pair of skis, a bike and an accordion to learn to play. Although Culhane's technique may be a little odd, she believes that children have to be kept busy to keep them out of trouble.

Instead of increasing the lengths of sentences for young offenders, the 74-year-old founder of the Vancouver-based Prisoners Rights Group suggests increasing the amount of community funding. Culhane stresses that no offenders, whether youth or adult, are treated fairly, since many of the people in prison are poor.

"Who are we (society) kidding?" she asks. "We're not looking for justice, we're looking for revenge."

Her solution: Move the money from prisons to daycare, invest in kids before they become a problem.

Bob Boden echoes her concerns. Formerly employed in a secure custody facility for young offenders, and now an attendance counsellor with the school system, Boden believes the act has strengths and weaknesses.

People look to the act for social miracles. "We err when we think that if we change it, it's going to bring about the changes in society that are missing."

He recommends a second look before blaming the system. "When crimes happen by young people in our community, we've got to say — as a community — what are we doing? We can't say the YOA failed, we can't say the JDA failed. We can't say the corrections system failed. We have to ask ourselves what we can do to help these kids. I've got to ask myself."

Parents of both victims and offenders are key players in this drama, Boden says. "There are some horrendous parents," he stresses.

Let's move from punishment to compassion: that's Boden's message. "We feel a lot of vengeance, but we don't spend a lot of time tolerating and loving and forgiving and understanding."

When people say the centres are like a Holiday Inn, Boden paints a picture of life at the "Holiday Inn."

"Say I sentence you to two years at the Holiday Inn. Now you can have all the food you want, you can watch all the TV you want, you can use the gym but you cannot leave the grounds for two years. How would you feel? Wouldn't that become a jail?"

Y.O.A. STATS

- 75% of the cases in youth court in 1990–91 resulted in a guilty verdict, according to a Youth Court Survey for the Canada Centre for Justice.

- probation was the most serious sentence, used in almost half the cases.

- 95% of offenders in cases involving murder and manslaughter were placed in custody.

- from '86–'91 short term sentences have increased while long term sentences have decreased.

- more than half the "guilty verdict" cases involve offenders in the 16–17 age range. 15% of these offenders end up in secure custody while younger offenders receive probation orders.

1982 BEFORE THE YOA

Teenagers (ages 12–18) made up 10.3% of Canada's population. 6.1% of them were convicted of crimes, more than 89 000 youths in total.

1986 AFTER THE YOA

Teenagers made up 8.8% of the country's total population, with 6.3% of them committing offenses, more than 139 000 in total.

1991

Teens comprised 8.1% of population, with 7.6% of them committing crimes, more than 171 000 in all.

Take a look. Despite cries that crime is rampant with the YOA, the figures don't indicate that at all. But, it hasn't dropped, either.

Welcome to
Reform School

♦♦♦

BY WALN K. BROWN

In his autobiography The Other Side of Delinquency, *from which the following excerpt is taken, Waln Brown describes his teenage years as a delinquent in the 1960s and his struggle to escape from the cycle of crime and institutionalization. After years of trouble at home, at school, at work, and with the justice system, Brown eventually fulfilled his desire for a college education and earned a Ph.D., with a dissertation on gang delinquency. He continues to work in the field of juvenile justice.*

January 13, 1961, was a gray, bitter-cold day. Snow and ice crunched under foot as I was led from the detention home and put in the rear seat of a car. Handcuffs bound my wrists. A metal screen separated me from the two policemen who had been given the duty of transporting me the two hundred and fifty miles [four hundred kilometres] to the Pennsylvania Junior Republic. Few words were exchanged. The cops were not pleased with their assignment. I was pondering the unknown perils of reform school. It was mid-afternoon when we drove between the brick pillars that defined the entrance to the PJR. The narrow macadam road wound past a number of old buildings. Double lines of boys marched past us, hunched against the chill winter wind. We drove halfway around a large oval, then stopped before a long white building marked "Administration." The two cops sandwiched me between them and headed through the double doors into the lobby of the building.

For the next hour I was moved from one office to the next and asked question after question. Then, when it seemed I had run out of answers, I was introduced to a man named Uncle John, the houseparent at the Inn Cottage and supervisor of the orientation detail, both of which I had been assigned to, and he escorted me back out the double doors of the building where nearly twenty boys stood waiting at attention. After placing me at the rear of the double column, he shouted some orders, and off we marched, moving among the buildings, picking up and letting off boys, getting physical examinations and hair cuts, taking tests, and requisitioning clothing.

It was nearly supper time when we marched into the cottage basement that was alive with a confusion of activities. Boys were everywhere wrestling, smoking, talking, shouting, singing, or dancing to the sound of a radio that blared in the background. The noise slackened as soon as Uncle John was sighted. Though a short man who looked to be in his fifties, he had a bulldog build that demanded respect. His eyes darting among the faces of the basement occupants, he gestured to a tall, dark-haired boy, told him to find me a locker, then walked away as though I no longer existed. I stood awkwardly by, waiting for the boy's instructions. He motioned me to follow him, pointed to one of the open-faced wooden lockers that was empty then he, too, left. Frightened, uncertain, I looked for a place to hide.

Before I could get the box of clothes given to me by the institution's tailor shop put away in the locker, the hazing started. Cigarettes were the first thing asked for or taken. Requests and demands to do laundry and other chores soon followed. Name-calling also began: crater-face, pimple-pusher. I was surrounded, engulfed in a circle of pushing, taunting, threatening boys. I wanted to cry, to strike out, to flee, but could only stand rigidly fixed to a spot, red-faced and confused, wishing it would stop. Suddenly the circle parted and two boys strolled through the crowd to where I stood. The others grew silent. I reached out a hand in a gesture of thanks for their help. They ignored it. Instead, they picked through my belongings, took what they wanted, then left, laughing, while the other boys looked on and grumbled. With little left worth taking, the crowd quickly dissolved. I took a seat on the long wooden bench that ran in front of the lockers and stared forlornly through moist eyes. Never before had I felt so humiliated, so defenseless. I wanted to die.

There was no time for self-pity. A "come and get it," followed by the thunder of hungry boys ascending stairs, announced the evening meal. I brought up the rear of the line. At the top of the stairs, I was turned back. Shoes were not allowed upstairs, only slippered and stocking feet. I rushed back, rummaged through the tailor-shop box, pulled on the striped slippers, and returned to the top of the stairs, which led to a dining room where more than fifty boys stood behind chairs at the small square tables. I grabbed a spot and followed the lead of the other boys who were staring toward Uncle John and his wife, Aunt Emma.

"Announcements!" barked Uncle John. The room went quiet. "The Inn basketball team plays Main Hall tonight at 7:15. Choir practice will be from 7:00 to 8:15 in the administration building. Sproul Hall has been canceled for the rest of the week. Hospital line: Daniels, Baxter, Smith, Harris and Yerkovitz. Smitty, you take the hospital line. Let's bow our heads."

"Our Father, who art in heaven…" the voices blended.

The prayer's end brought an explosion of noise. Tables and chairs slid along the floor, dishes and silverware collided, and unfinished conversations were renewed.

"Quiet!" boomed Uncle John.

The noise dropped to a dull roar.

Fifteen square tables filled the dining room. Three or four boys arranged themselves at each table. Boys wearing white streamed between the kitchen and the tables, bearing pitchers of milk, pots of soup, and loaves of bread. One boy served Uncle John's table, where meat and vegetables took priority over soup. Occasionally, Aunt Emma threw instructions at those who served. The two boys who had picked through my belongings and the tall boy who had shown me to my locker sat at the table nearest Uncle John. Uncle John handed them the leftovers from his table. I ate very little.

We were dismissed from the meal. A mob of boys made a mad dash for the basement. I followed less enthusiastically.

The basement hummed with activity as I slunk down the stairs toward my locker. A tall blonde kid threw a left hook at the head of a smaller boy. The boy fell to the bench at his rear. I took a seat, lit my remaining cigarette, and pretended interest in my hands to help hide the fear.

No sooner had I finished the last puff of the cigarette than someone from upstairs yelled "shineline." A number of boys filtered through the crowd, picking boys and sending them upstairs. I was among the group selected.

Dozens of mock-skaters slid across the dining-room floor, pieces of cloth under their slippered feet as they glided silently across the wooden planks, pivoting at the walls, then resuming their quiet skate. I stepped onto two rags and imitated the others. We slid back and forth across the room until the floor shone beneath our feet.

Again we were dismissed to the basement. I sat by the locker, quietly, trying to blend into the woodwork, while other boys fought and played. Some took their evening showers in the large cement stall that was open to full view. Others changed into pajamas and nightshirts. I found a nightshirt in the tailor-shop box and quickly threw it on, afraid to expose my pimply body to their mockery.

"Line up! Let's go! Put them cigarettes out and get in line!" commanded the tall, dark-haired boy as he and Uncle John entered the basement.

Boys stopped what they were doing and formed a line along the lockers. Uncle John and the tall boy moved along the line, counting noses.

"Forty-six," announced Uncle John.

"Forty-six," parroted the tall boy.

"Count's right," confirmed Uncle John. "Forty-six present, five on late hospital line, and one AWOL."

Uncle John scanned the line of boys. He said nothing but appeared to be singling out certain boys with his gaze. The basement was uncomfortably silent.

"There will be no disturbances like last night," scolded Uncle John....Now bow your heads for the Lord's Prayer."

"Our Father, who art in heaven..."

I shadowed the flock of gowns up the three flights of stairs that led to the bedrooms. The tall boy grabbed my arm and guided me toward a bunk. I crawled onto the top one and rolled beneath the covers. The lights were put to rest, but my mind was not.

I was awake most of the night, sleep coming only in the wee hours of the morning and then quickly put to flight by the demons of my dreams. Bedsheets torn apart by the night's turmoil were draped in disarray over the side of the bunk. I lay in a cold, shivering ball cursing the invasion of the January sun.

"Rise and shine!" a voice echoed from another room.

I rolled onto my side, turning away from the announcement of another day to be endured.

"Get up and shine!" demanded the voice again, closer.

Around me, bodies began to stir. Springs groaned, limbs stretched and mouths gave vent to yawns. Reluctantly, I dropped to the floor and made my bed in the military fashion displayed by the others. Some already trudged the wooden floor, shinerags beneath their listless feet. Like slow-motion robots they skated, as though clinging to a few cherished moments of tranquillity. For ten minutes we slid squinty-eyed around the room until commanded to quit our efforts and prepare for breakfast.

Following the morning meal most of the boys flocked through the basement door, heading for work or school. Those of us on orientation remained behind. We were not yet allowed to do as the others. We awaited Uncle John's command.

"Get outside and form ranks!" barked Uncle John.

We filed through the door and made a double line.

"Attention!"

Bodies snapped rigid.

"Dress right, dress!"

Arms reached sideways to touch a neighbor's shoulder. Bodies rhythmically adjusted position.

"Eyes front!"

Falling hands slapped thighs. Heads snapped forward.

"Right face!"

Bodies quarter-turned with varying degrees of smartness. With each order, I attempted to copy the moves of those around me.

"Forward march! Your left. Your left. Your left, right, left. Keep in step, Brown."

The day was spent marching from one place to another, learning the layout of the campus. Nearly three hundred fifty boys inhabited seven cottages. A farm, tailor shop, chapel, guidance office, gymnasium, recreational building, shop, swimming pool, playing field, and a combination administration/school building rounded out the grounds.

Sometimes we worked at various group assignments: shoveling snow, performing farm tasks, picking up litter, cleaning buildings and rooms. Around such daily chores, boys were constantly being dropped off and picked up from scheduled events: hair cuts, counseling sessions, tests, medical appointments. But most of all we marched to the commands of Uncle John and learned to follow orders. On Saturday evenings there was a movie. Sunday mornings meant Church. Sunday afternoon and evening was unstructured, leaving us to fend for ourselves in the basement.

For twenty-eight days I was subjected to the same routine. Then, just as I was getting used to it, Uncle John called me to his room and informed me that I was to be transferred to another cottage. Though I disliked orientation, and was fearful of many of the boys, I had made a few friends and did not wish to begin from scratch. My pleas to remain at the Inn were denied. Two days later, I left.

Doing Time at Kirkwood

♦♦♦

BY

DANA

WHITE

Kirkwood ski resort is only 150 miles [240 km] from San Francisco Bay. But for Nick, a 16-year-old who has been arrested twice for dealing heroin in Oakland, it might as well be a different planet.

"At first I was too scared to go on a lift; I'm afraid of heights," says the powerfully built young man, his face engulfed by a pair of orange Oakley goggles. "Now that it's cool, I'm not scared no more."

That I came to be sharing a chair lift at Kirkwood in mid-April with an ex-drug dealer resulted from an interview I'd had three weeks earlier with extreme skier Glen Plake. Plake had mentioned his involvement with Rite of Passage (ROP), an innovative program that uses athletics — including skiing — to rehabilitate troubled youths. An old ski buddy of his from Lake Tahoe named Ski (yes, it's on his birth certificate) Broman had been instrumental in its formation.

ROP takes male juvenile offenders, ages 13 through 17 from 26 counties throughout California, who have been arrested for nonviolent crimes or for violating probation. Most of them are high school dropouts and "multiple placement failures" who have exhausted more traditional avenues of reform in their community or county. ROP is often their last chance to straighten out before being shipped off to the California Youth Authority prison for juveniles and, potentially, lost for good in the black hole of the criminal justice system.

"Skiing always kept me out of trouble, always gave me a focus." At 25, Ski Broman, part-owner and executive vice president of ROP, isn't much older than the teenagers he works to help. A Tahoe native, he spent his teens skiing the South Shore resorts and racing in high school and on the Far West Ski Association circuit. At 17, driven by his belief that athletics can change behavior, he and several similarly motivated friends started a group home that used athletics as a "treatment modality." Shortly thereafter he left to pursue his ski racing career at the Mission Ridge Ski Academy in Wenatchee, Washington. He didn't become a world-class competitor, but his experience at the academy did influence him to "make something of myself and to do something worthwhile." He returned to Tahoe and helped make ROP a success.

Skiing is only one of many athletic pursuits offered by ROP, but it's the sport that's closest to Broman's heart. He feels it has something unique to offer.

"For these kids, being exposed to skiing and to the environment is both physically challenging and intellectually refreshing because it allows them to interact with people that have made a commitment to a lifestyle that is healthy and positive," says Broman, who is pursuing a master's degree in marriage, family, and child counseling at the University of San Francisco. "Plus it surrounds them with an environment that is a lot different from the concrete streets that these kids come from. Just being up there on the top of the mountain is an accomplishment for a lot of these kids and an experience that they would never have had without coming to ROP."

Of the 200 youths who enter ROP every year, only about 80 ultimately ski. By the time a boy gets to that point, he's spent about nine months working his way through two previous phases of the program. It's no stroll in the park, and about 40 percent of the participants drop out.

The journey to Kirkwood starts at ROP's headquarters in Minden, Nevada, which lies near the California border, about 30 miles [60 km] from Kirkwood. Fresh from a county juvenile detention facility, a new arrival — often handcuffed — is processed here and then transported to the aptly named Remote Training Center (RTC), an austere outpost of whitewashed buildings and Quonset huts on a dry lake bed in the Nevada desert. Here he goes through several months of rigorous workouts and classroom study.

If he makes it through the RTC, he moves on to the Athletic Training Center (ATC). Here he continues his education and receives intensive counseling and treatment for issues like domestic violence and alcohol and drug dependency. He can choose to participate in three of 12 team sports at the ATC, competing against local division-A high schools. (Last year [1990] the ROP Raiders were the state division-A champs in basketball and football.)

The opportunity to ski comes at the final phase of the program: 12 weeks in a qualifying house, known as a "Q house," where the focus is on individual sports. He can choose to live in one of two Q houses — Q1, in Minden, offers tennis and mountaineering in the summer, Alpine skiing in the winter, and lifeguard training year-round; Q2, in Carson City, Nevada, offers mountaineering, rock climbing, and Nordic skiing. The boys also do extensive community service work, including various tasks at Kirkwood in exchange for discounted season passes.

"The sports at the Q houses," Broman explains, "are an introduction to a lifestyle. We want these kids to utilize their free time doing quality things, rather than antisocial things, as they have in the past. Rip, run, and steal is what they did during their free time. You replace that with bicycle riding and mountain climbing, and you're doing everyone a big favor."

By the time I get to ski with the boys at Kirkwood, the snow is fading fast. On a sunny Thursday afternoon, nine boys and four coaches line up at the base, where Broman and his head ski instructor, Jon VanDeventer, lead them through stretching exercises. After a series of on-slope drills — including the "human slalom" — the kids are divided into groups according to ability. There are certain ground rules when on the hill. They can ride the lifts only with other Q-house skiers. Once at the top, they must line up, and then ski down as a group. In their uniforms — powder-blue bib-pants and red jackets accented with blue and white stripes — they're hard to miss.

"People come up and ask, 'Are you a ski team?'" comments VanDeventer. "And we say, 'Yeah, we're a ski team. As a matter of fact, we have a few dry-land training centers.'"

All day I tail the more advanced group around the mountain. Broman, a fantastic skier, leads the way down single- and double-diamond runs — cornices, chutes, slush bumps — that make me hesitate but that the boys attack with a defiant pluck that's both poignant and alarming at the same time.

"We're into positive reinforcement here," said VanDeventer. He and his staff instruct with a light hand to fit the emotional needs of boys who are on a first-name basis with failure. "These kids don't take criticism real well. But generally, they're ideal students as far as someone who's physically fit and aggressive. You can show them things that you wouldn't be able to show a normal person on their fourth day of skiing. It just blows my mind, the progress a lot of them make."

At one point Broman and I run into Mike, an ROP alumnus with a wild mop of dark blond curls. Mike, who goes by the nickname "Wigger" ("'cause I wig out," he explains) confides that his parents beat him, and that at 13 he got caught in a revolving door of truancy, drugs, theft, and arrest. He first skied in ROP, loved it, and after leaving the program got a paying job at Kirkwood, moving up from parking cars to working in the rental shop. He skies almost every day; it helps give his life a sense of balance.

"The one thing I like about skiing is that it's a form of art, I believe," Mike says, waggling one of his 205-cm Head GS skis, the tip of which is adorned with a Grateful Dead sticker. "I can make turns and shred hard. I really love it when people go, 'Wow, that guy can do that really well.'"

Mike is still struggling to stay clean and hold his life together; it's tough going sometimes, but it beats the alternative: "I could've been doin' time in jail. Now I'm doin' time on the slopes. Skiing's in my blood. I can't stop skiing."

Like the extreme skiers they admire — Scot Schmidt, Plake — these boys crave the rush of skiing on the edge. What do they prefer? "Jumping." "Big air." "Cliffs." So I'm not surprised when later that afternoon the group convenes beside a gully with a lip on the far side that looks like a promising launching point. One boy worries that he'll fly too high. A colleague with a closely shaven head mumbles that it's no big deal — he used to jump from second-story windows with a VCR under each arm. Each boy takes the jump in turn, with varying degrees of grace. Mike catches the biggest air and throws in a stylish daffy to boot.

The next day is the Q houses' last ski day of the season, and to mark the occasion, coaches and counselors from the RTC and the ATC are invited to ski with the kids. Summer seems upon us; the temperature exceeds 70 degrees [21 C] . It's Friday the 13th, but no one seems to notice — there's a barbecue in the parking lot, and at lunch Broman's staff surprises him with a plaque of appreciation for his years of effort and a white-frosted cake sculpted in the shape of a mountain topped by a tiny plastic skier.

The picnic gives me the chance to speak with Nordic skiers from the Q2 house. I'm told that the best skier is Gabriel — he can do garland turns on telemark skis — but he's too shy to talk. Taciturnity isn't a problem with Robert and Kevin, two street-gang members who have become fast friends. Everyone else has peeled down to T-shirts in the early heat, but these two are bundled to the teeth. Wearing identical blue stocking caps, blue shells, and leather telemark boots with blue laces, they look like matching bookends. I milk them for their opinions about skiing, but they don't seem overly impressed by it, and keep steering the conversation to girls instead.

If this were a B-movie script, every Q-house skier would, like Mike, fall in love with the sport and attempt to shape a life around it. The fact is, many of these boys will never ski again. But hopefully the sensation will stay with them, a window opened on a brighter world. I recall riding a lift with Maso, a slight boy with black hair and a fledgling moustache. "Skiing's a bigger thrill than dealing drugs," he had remarked, taking in the thawing landscape as if to commit it to memory. "This is the better side of life."

Note: Nick now lives in a Christian group home; he played high school football last fall. Maso is living on his own, working in construction, and still skiing. Robert lives in a group home and attends public high school. Kevin's chronic truancy from school keeps him in and out of rehab.

An official ROP ski team is competing against local high schools this season.

Elders' Sentences Help Keep Youth in Line

♦♦♦

BY

JOE

McWILLIAMS

When the youth of Wabasca-Desmarais break the law, they get charged and appear in court just like anywhere else.

And just like anywhere else, if they plead guilty, they get what's coming to them. But since Feb. 1991, it hasn't been a judge who metes out the punishment.

Five leaders of the largely Native community northeast of Slave Lake have been handling that job. And according to all reports, whatever they're doing is working.

"Everyone concerned believes that it's because of the Youth Justice Committee," said Raymond Yellowknee, an adviser to the elders' group.

The way for this type of community-based system of justice was paved by Provincial Court Judge Clayton Spence. Spence is committed to accepting the committee's sentencing recommendations, because he knows what came before wasn't working.

There was a serious lack of respect for the decisions of the court among young offenders, much as there is in many communities. That trend seems to have been reversed with the elders now playing the role of judge.

Yellowknee said a key component of the program is reconnecting the youth with traditional Native practices and values. Respect for elders, especially grandparents, was a cornerstone of traditional aboriginal society, he said. And he sees that dynamic coming into play in the relationship between the elders and offenders.

"My feeling is the youth never considered what they were doing was wrong. When they have to speak up to the elders and justify their reasons, they can't do it."

Each youth and at least one of his or her parents attends an interview with the committee. The elders question the offender's motivation and also probe the relationship with his or her father and grandfather. The reason, said Yellowknee, is that many youths don't have one and the lack of connection to paternal authority is one of the main causes of bad behavior.

So they may sentence a youth to work for or with a grandparent chopping wood, hunting or repairing fences. Orders to apologize to the victims of assaults or property offences are also often handed down, said Yellowknee.

One youth was given an order to apologize and do some work for the same man.

"I saw him later and he came running up to me and told me, 'I apologized to that guy yesterday!' It seemed like one of the few positive things that had happened in his life."

Sentences to spend time with an older relative hunting or trapping are also popular.

"They're trying to give them a look at what life was like years ago. I believe we'll see more of that," said Yellowknee.

That may sound soft, but the committee is not afraid to recommend that a youth be sent away for alcohol treatment, if they think other measures won't work. Judge Spence thinks the committee's recommendations are appropriate. And, as he points out, crime is down, and that's what counts.

The success of the program has not gone unnoticed. So far, five delegations from other Alberta communities have come to Wabasca for first-hand information. Two members of the Alexander reserve west of Edmonton came and sat in on one of the elders' offender interviews, "a big plus for them," Yellowknee said.

Groups from Morley, Hobbema, High Level and Lac La Biche are also considering following Wabasca's example. Spence has heard from judges in other communities, asking how the program works. Crime in Fort Chipewyan is "down across the board" in the year-and-a-half since a community group took over sentencing young offenders there. In Anzac, near Fort McMurray, a community-based sentencing group has been up and running for about four months now.

Yellowknee sees a trend developing in Native communities across the country, hand-in-hand with the movement toward Native self-government. Justice by elders' committee is the answer, he believes, not having more Native judges.

"I think if we keep going like this, the next step will be to handle all cases," he said.

The Practice of Mercy

♦♦♦

BY

**ALDEN
NOWLAN**

Beginning the practice of mercy,
study first to forgive
those who have wronged you.

Having done that,
you will be ready
for the sterner discipline:

learning to forgive those
you have betrayed and cheated.

ACKNOWLEDGEMENTS

Care has been taken to trace ownership of copyright material contained in this text. The publishers will gladly accept any information that will enable them to rectify any reference or credit in subsequent editions.

TEXT

p. 1 "This Is a Law" by F.R. Scott. From *The Collected Poems of F.R. Scott* by F.R. Scott. Used by permission of the Canadian Publishers, McClelland & Stewart, Toronto; **p. 8** "Groceries" by Monty Reid from *The Life of Ryley* (Thistledown Press Ltd., 1981), used with permission; **p. 9** "Guilt" by Leona Gom. From *Private Properties* by Leona Gom, 1986. Reprinted by permission of Sono Nis Press; **p. 10** "Win, Lose or Learn" by Peg Kehret. From *Acting Natural* by Peg Kehret. Copyright © 1991 Meriwether Publishing Ltd., 885 Elkton Drive, Colorado Springs, CO 80907 USA. Used by permission; **p. 13** "The Mouthorgan Boys" by James Berry from *A Thief in the Village*, copyright © James Berry, 1987. First published by Hamish Hamilton Children's Books, 1987. Reproduced by permission of Hamish Hamilton Ltd.; **p. 18** "Maria Preosi" from *Class Dismissed!* by Mel Glenn. Copyright © 1982 by Mel Glenn. Reprinted by permission of Clarion Books/Houghton Mifflin Co. All rights reserved; **p. 19** "The School Yard Bully" by Peg Kehret. From *Encore! More Winning Monologs for Young Actors* by Peg Kehret. Copyright © 1988 Meriwether Publishing Ltd., 885 Elkton Dr., Colorado Springs, CO 80907, USA; **p. 23** "Blessed Are The Meek" by A.P. Campbell first appeared in *Canadian Review* in 1975. Reprinted by permission of the estate of A.P. Campbell; **p. 27** "Oranges" by Gary Soto. From *New and Selected Poems* by Gary Soto (Chronicle Books, 1995); **p. 29** "Miss Calista's Peppermint Bottle" by L.M. Montgomery. From *Among the Shadows* by L.M. Montgomery. Used by permission of the Canadian Publishers, McClelland & Stewart, Toronto; **p. 35** "Becoming a Judge Vindicates her Father" by Judy Steed. Reprinted with permission — The Toronto Star Syndicate; **p. 42** "Slocan Diary" by Kaoru Ikeda. From "Slocan Diary" by Kaoru Ikeda in *Stone Voices: Wartime Writings of Japanese Canadian Issei* edited by by Keibo Oiwa. Reprinted by permission of Vehicule Press; **p. 47** "That Summer I Left Childhood Was White" by Audre Lorde. Excerpt from *A New Spelling of My Name* by Audre Lorde, The Crossing Press, Copyright 1982; **p. 52** "Question and Answer" by Langston Hughes. From *The Panther and the Lash* by Langston Hughes. Copyright © 1967 by Langston Hughes. Reprinted by permission of Alfred A. Knopf, Inc.; **p. 53** Excerpt from "Where the Rivers Meet" by Don Sawyer. From *Where the Rivers Meet* by Don Sawyer. Winnipeg: Pemmican Publications, Inc., 1988. Reprinted by permission; **p. 69** "Speaking the Language of the Soul" by Mary Lou Fox. By permission of Mary Lou Fox, Anishnabe Kwe, Wikwemikong First Nation, Manitoulin Island, Ontario; **p. 73** "Statement of Vision Toward the Next 500 Years" by Artists, Writers and Wisdom Keepers Gathering in Taos, Home of the Red Willow People, Co-Chaired by Suzan Shown Harjo and Oren Lyons. Reprinted by permission of The Morning Star Institute, Washington D.C.; **p. 76** "Agnes Macphail: Reformer" by Doris Pennington. Copyright © 1989 by Doris Pennington. From the book *Agnes Macphail: Reformer*, published by Simon & Pierre Publishing Co. Ltd. Used by permission; **p. 82** "Ascendancy" by Sheila Deane. The poem originally appeared in *Canadian Woman Studies/les cahiers de la femme*, "Women in Science and Technology: The Legacy of Margaret Benston," Volume 13, Number 2 (Winter 1993); **p. 85** "Women Pay More for Identical Items" by Charlotte Parsons. By permission of *The Globe and Mail*. **p. 89** "On

Aging" by Maya Angelou. From *And Still I Rise* by Maya Angelou. Copyright © 1978 by Maya Angelou. Reprinted by permission of Random House Inc.; **p. 90** "Society Discriminates Against Teenagers" by Kathleen Cawsey. Reprinted by permission of Kathleen Cawsey; **p. 92** "Kingship" by Kit Garbett. Reprinted from *New Internationalist, The Monthly Magazine on World Justice and Development Issues*, Tel. (416) 588-6478; **p. 95** "The Convention on the Rights of the Child." By permission of the United Nations Centre for Human Rights and UNICEF. Reprinted from the *Unesco Courier,* October 1991; **p. 104** "A Small Crime" by Jerry Wexler. Copyright © Jerry Wexler. "A Small Crime" was previously published by *The Montreal Review, Jewish Dialogue Magazine, Worlds Apart* (Cormorant Books) and in the collection *The Bequest and Other Stories* by Jerry Wexler (Vehicule Press); **p. 106** "Plenty" by Jean Little. From *When the Pie Was Opened* by Jean Little. Copyright © 1968 by Jean Little. By permission of Little, Brown and Company; **p. 108** "Walking Tall, Talking Tough" by Fiona Sunquist. Copyright National Wildlife Federation. Reprinted with permission from *International Wildlife*, Sept./Oct. 1991 issue; **p. 114** "The Thief Who Kept His Hands Clean" by Carol Kendall and Yao-wen Li. From *Sweet and Sour: Tales From China* retold by Carol Kendall and Yao-wen Li, 1978. Reprinted with permission from The Bodley Head Ltd.; **p. 116** "The Orangutan Who Believed In Simple Solutions" by Alan Neidle. Reprinted from the book *Fables For the Nuclear Age* by Alan Neidle. Copyright 1989 by Paragon House Publishers. Reprinted with permission of Paragon House Publishers; **p. 118** "What's the Verdict?" by Ted LeValliant and Marcel Theroux. Used by permission of Sterling Publishing Co., Inc., 387 Park Ave. S., NY, NY 10016. From *What's the Verdict?* by Ted LeValliant & Marcel Theroux, © 1991 by Ted LeValliant & Marcel Theroux; **p. 123** "Court Jesters" by Peter V. MacDonald. From *More Court Jesters* by Peter V. MacDonald. Reprinted with the permission of Stoddart Publishing, Don Mills, Ontario, Canada; **p. 127** "Partners in Crime Prevention" by Luis Millan. Reprinted with the permission of *Canada and the World* magazine, Waterloo, Ontario; **p. 130** "My Theme" by Hans Jewinski. Courtesy of Hans Jewinski; **p. 132** "Young Offenders Act: Pros & Cons" by Lori Martin. Reprinted by permission of *TG Magazine*; **p. 136** "Welcome to Reform School" by Waln K. Brown. From *The Other Side of Delinquency* by Waln K. Brown. Copyright © 1983 by Rutgers, the State University. Reprinted by permission of Rutgers University Press; **p. 141** "Doing Time at Kirkwood" by Dana White originally appeared in the March 1991 issue of *Skiing* magazine; **p. 147** "Elders' Sentences Help Keep Youth in Line" by Joe McWilliams. Story by Joe McWilliams for *Windspeaker*; **p. 149** "The Practice of Mercy" by Alden Nowlan. From *An Exchange of Gifts* by Alden Nowlan. Reprinted with the permission of Stoddart Publishing, Don Mills, Ontario.

PHOTOGRAPHS

pp. 3, 20 Dick Hemingway; **p. 36** The Toronto Star/M. Stuparyk; **p. 46** UPI/Bettmann; **p. 51** UPI/Bettmann; **p. 55** Dorothy Chocolate/Native Communications Society of the Western NWT/Native Indian/Inuit Photographers' Association (NIIPA); **p. 65** Woodland Cultural Centre; **p. 68** Joseph Shebagegit Jr./NIIPA; **p. 72** Camela Pappan/NIIPA; **p. 77** UPI/Bettmann; **p. 84** left: Glenbow Museum NA-11514-3, centre: National Archives of Canada/NFB PA-143958, right: Canapress Photo Service; **p. 97** UNICEF/Ruby Mera; **p. 103** UNICEF/Mainichi/Shinichi Asabe; **p. 113** Fiona C. Sunquist; **p. 122** The Bettmann Archive; **p. 126** Canada Wide Feature Service Ltd.; **p. 146** Woodland Cultural Centre/Greg Staats.